Asya Schween

MICHAEL PAUL MASON

Head Cases

Michael Paul Mason, born in 1971, is a brain injury case manager based in Tulsa, Oklahoma. He writes for *Discover* magazine.

HEAD CASES

HEAD CASES

Stories of Brain Injury and Its Aftermath

MICHAEL PAUL MASON

FARRAR, STRAUS AND GIROUX
NEW YORK

Farrar, Straus and Giroux
18 West 18th Street, New York 10011

Distributed in Canada by Douglas & McIntyre Ltd.
Printed in the United States of America
Published in 2008 by Farrar, Straus and Giroux
First paperback edition, 2009

The Library of Congress has cataloged the hardcover edition as follows:
Mason, Michael Paul, 1971–
Head cases : stories of brain injury and its aftermath / Michael Paul Mason.—1st ed.
p. cm.
Includes bibliographical references.
ISBN-13: 978-0-374-13452-5 (hardcover : alk. paper)
ISBN-10: 0-374-13452-9 (hardcover : alk. paper)
1. Brain—Wounds and injuries—United States—Case studies.
2. Brain—Wounds and injuries—Patients—Rehabilitation—United States—Case studies. I. Title. II. Title: Stories of brain injury and its aftermath.
[DNLM: 1. Brain Injuries—United States—Personal Narratives.
2. Health Services Accessibility—United States. 3. Long-Term Care—United States. WL 354 M398h 2008]

RC387.5.M377 2008
617.4'810443—dc22

2007032335

Paperback ISBN-13: 978-0-374-53195-9
Paperback ISBN-10: 0-374-53195-1

Designed by Debbie Glasserman

www.fsgbooks.com

1 3 5 7 9 10 8 6 4 2

Some names and identifying details have
been changed to protect privacy.

FOR THE INJURED

CONTENTS

HEAD CASES

INTRODUCTION

The first thing I tell her is that I cannot help. Her son Jake is thirty-four, my age. His gray, bruise-flecked limbs are splayed out on a bed before me; his mouth is dry and agape. I know I cannot help him. I cannot file a lawsuit against the insurance company, I cannot conjure a way out of this dead-end nursing home, and I cannot sucker punch the aloof neurologist or throttle the ignorant psychiatrist. I hold no sway over the waiting list in my own hospital. I explain to her that I can do nothing at all, and she sighs. She is desperate to see Jake in a program where there is a sense of progress and direction. She knows that the rehabs and specialty hospitals are as inaccessible as the moon. She has called them herself, and she knows that nobody can help. She knows I cannot help, but she asks me anyway. She asks, in all earnestness, to do the impossible and find her son a bed, and in my weakness, I agree. It's my job to agree.

Jake turns his head toward me and I suspect he can hear me. If he can respond, no matter how minimally, then he meets the most important criteria. He closes and opens his mouth arbitrarily, but not a sound comes out. I ask him to lift his head and I wait. Nothing. His mother is quiet and nearly in tears. I ask her to turn off the fluorescent lights and shut the door, and when she does Jake exhales faintly. It sounds like relief, like the hint of a response. I ask Jake once more to lift his head. A good fifteen seconds later, his head slowly raises an inch off the pillow, and then drops back down. That's criteria enough for me. His mother grins at me proudly, as though her son just won a marathon. In a sense, he has.

The next two hours find me thumbing through a thick notebook of Jake's medical records, trying to decipher the scribbled progress notes and lab reports, then interviewing nurses and aides and doctors. I spend the last half hour of my evaluation explaining to Jake's mother that this will take months at a minimum, but most likely a year or two—if anything happens at all. She will be my best resource through all of this, I tell her, as I hand her a list of administrators to call and a few forms to fill out. She tells me she will do this and anything else I ask, but she doesn't need to say a word. If she's been through this much already, paperwork and phone calls aren't going to stop her.

Before I leave, she asks for my business card, not because she doesn't already have my number, but because she wants to know my title, she wants to know what kind of person drops everything to go to look at a bedridden man four hundred miles away. I hand her my card and tell her I am a brain injury case manager, and there is the hint of a smirk because we both know that brain injuries cannot be managed any more than a thunderstorm can be managed. They can be endured, accepted, and integrated, but not managed. She clasps her hands around mine, and I say goodbye, and I get back in my car. I am forever back in my car. I turn the ignition and hope for home, so I can lie down next to my wife and rest my hand on the warm crest below her navel.

The hope is false; a singular voice mail from my director asks me if I can drive two hours north to Sioux Falls to assess a woman with brain damage due to a viral infection. Either I drive the two hours today, or I drive back home to Tulsa and drive ten hours up to South Dakota on Monday. I head north and can already hear the disappointment in Christy's voice. When I call her from a filling station outside of Omaha, she doesn't answer and I leave her a message. My marriage unfolds in messages.

On my way up to Sioux Falls, I think about Jake still lying there in the nursing home, and how he has been lying there throughout the past year while I was buying a house and mowing the lawn and attending parties and traveling. He has been lying there long enough for his wife to have given up hope and file for divorce, and he hasn't seen his son in months. Jake is trapped in an awful sense of nowhere, and yet he is still present, still responding, still human. I may not be able to help, but I can act. I can act for Jake, which, right now, is more than he can do for himself.

I don't look for cases; the cases find me. They catch me in airport terminals and waiting rooms, they crowd my inbox and voice mail. A single phone call can send me to Wyoming one day and Indiana the next. During a tough week, I might wind up in a dozen different cities. When I first started traveling, I would forget what city I was in when I woke up, so I would leave a phone book by the nightstand to orient me in the morning. I still forget the city I'm in, but now I leave the phone book in the drawer. The city comes to me sooner or later. Where I am at doesn't matter so much; it's the catastrophe that sets me in motion.

Medicine reserves the term *catastrophic* for a handful of conditions, and the term fits. Spinal cord injuries sever and parse a person's ability to move and control their limbs and basic body functions. Amputations disintegrate any hope of feeling and sensation. Full-body burns inflict a fierce pain only pharmaceutical amnesiacs can forgive. But chief among the catastrophic conditions is the traumatic brain injury, commonly called TBI. It is an upheaval of physical, psychological, social, and spiritual proportions.

The wry eighteenth-century haikuist Issa wrote, "In this world / We walk on the roof of hell / Gazing at the flowers." Those familiar with brain injury are familiar with Issa, if not by name, then in sentiment. Spend an hour in a room with someone who struggles to maintain eye contact with you and whose limbs are flailing as a result of a simple fall from a ladder, and soon enough a sense of your own vulnerability finds your legs less steady, your mouth a little dry, and your hand slow and cautious. Traumatic brain injury strikes with the concussive ferocity of a bomb; woe to those near its epicenter.

A tap on the head, and anything can go wrong. Anything usually does go wrong. You may not remember how to swallow. Or you may look at food and perspire instead of salivate and salivate when you hear your favorite song. You may not know your name, or you may think you're someone different every hour. Everyone you know and will ever know could become a stranger, including the face in the mirror. When you tell someone you're sad, you may shriek; your entire vocabulary may consist only of groans or hiccups. A brain injury can shatter your notions of the future, splinter your past, and send your sense of time whirling in any number of directions. And that's just the beginning.

A brain injury is never an isolated incident; it affects nearly everything associated with the survivor. It can collapse a family and flatten a business, evaporate friendships and allegiances, overburden a community, and buckle a state's healthcare system. Thanks to antiquated legislation and massive cutbacks, few states are capable of providing even the most minimal level of specialized care for brain injury patients, forcing most survivors to find treatment hundreds of miles from home, if they can find it at all.

Brain injury is a quiet crisis; the numbers are almost too large to make sense. As many as 5.3 million Americans are living with a permanent disability resulting from a brain injury, a full 2 percent of the population. The Centers for Disease Control and Prevention reports that 1.4 million Americans sustain a brain injury each year, and that fifty thousand people die from that injury. Almost a quarter-million people are hospitalized; the remaining number are treated and released from the emergency room. Some of the released go home comforted, only to discover that they no longer have a sense of smell or taste, or that their sleeping habits have changed, or that they can't seem to do their job anymore. There are more undiscovered head injuries in this world than are dreamt of in our medical journals. Only now are we beginning to understand something about the number of known brain injuries.

Reframed, the numbers nauseate. In America alone, so many people become permanently disabled from a brain injury that each decade they could fill a city the size of Detroit. Seven of these cities are filled already. A third of their citizens are under fourteen years of age. Currently, there are at least ninety thousand Americans with a brain injury so severe that it requires an extended stay in a post-acute brain injury rehab, but there are only a few thousand specialty beds, and upwards of 90 percent of them are already occupied. The severely brain injured are not getting the treatment they need—they're getting mistreated through neglect, misplacement, and isolation.

Numbers, however, are not lying in hospital beds, nor are they languishing in mental health asylums and prisons. A brain injury has a way of exposing humanity at its most vulnerable, fragile, and determined. Because the brain is composed of a million billion synaptic connections, each injury to the brain is as unique and complex as the life it affects. No matter how much detail a person's medical records indicate about their injury, the record is only a shadow, a small hint, at the human behind the injury. In the leveling world of TBIs, soccer moms grow heroic as soldiers, drug addicts transcend to the holy, the holy lose hope, and great egos give way.

The sequence of events goes something like this: the brain gets damaged, and two months later, the million-dollar insurance policy is depleted and the patient is shuffled out the door with a shrug of shoulders. A course of treatment that should have lasted years is cut short before it even starts. Bereft, the family starts mining libraries and websites and social service agencies for anything and anyone who can make sense, who can act. Meanwhile, the TBI survivor gets bumped from nursing home to psychiatric ward to emergency room to homeless shelter to group home and elsewhere. They subsist in a medical purgatory. For all their determination and perseverance, the caretakers and the patient wind up shortchanged with me. I show up as embarrassed as a fireman late for a fire.

Case managers number in the tens of thousands, brain injury case managers number in the tens of dozens. According to the Case Management Society of America, case management is a "collaborative process of assessment, planning, facilitation and advocacy," but the definition is genteel. Case managers are troublemakers and vigilantes. I became a brain injury case manager through a series of career miscalculations; I had accrued an uneven set of skills that included writing and mental health work, and found myself a misfit in most clinical settings. The job not only required a familiarity with medical facilities, but also demanded an exacting level of observation, one that could be translated into a comprehensive evaluation of a complex injury and eventually used as a powerful advocacy tool. My ears remain mine, but my voice belongs to the patient.

Of all the medically challenged injuries, traumatic brain injuries require the most involvement and cost over time. A hospital finance director in Boston once told me that he had never met a family that was financially prepared for the cost of a brain injury. It takes, on average, four million dollars over the course of a lifetime to treat a TBI. One person may require two medical caretakers around the clock, indefinitely. Another may benefit from a simple standard treatment lasting three months. But at a cost averaging a thousand dollars a day, most people can hardly endure the financial burden of one week's worth of rehabilitation, much less a stay that should last months or years.

. . .

The goal is to get a survivor unstuck, to move them from a stagnant setting into a life of greater independence. If the severity of their injury warrants it, their best chances for recovery lie in brain injury rehabilitation, a specialized form of care usually available only to the affluent or the tenacious. My job is to assess a person's needs, explore their available resources, and create a means for them to obtain the resource. The biggest obstacle I face is that there are usually no resources and no means available. Sometimes the only choice I have left is to pick a fight. A proper, legal, diplomatic fight, but a fight nevertheless.

After an initial contact—usually a phone call—I accumulate medical records and I review the case for validity. If a person doesn't have a documented brain injury, then the first step is to get it documented, usually through the help of lab reports and a neuropsychologist. Once the extent of the injury is measured, I review the records in detail, trying to enter the person's life as intimately as possible. Through further conversations with their family members or social worker, the rent images of the injured person collect and intensify in my mind. Within the course of days, I become an involved companion, though the survivor may not even know I exist. Once I feel familiar enough with the case, I start in with the phone calls to facility directors, admission coordinators, doctors, state officials, and attorneys.

The phone calls turn into e-mails, the e-mails into faxes, the faxes into plane tickets and gas pumps. The cycle from initial call to in-person evaluation can take two years or two hours, and still the evaluation is only a preliminary step. It occurs, mainly, to reconcile the survivor with their paperwork. I see for myself whether the person seeks help or has succumbed to hopelessness. I find out if their guardian or caretaker wants them better, or just wants them gone. Most important, I am there to listen. Much of what I am told in confidence never appears in a medical chart.

Over the years, I've found myself moving in the background of certain stories that challenge the very assumptions we make about being human. By going deeper into the aftermath of brain injury, we eventually reach an earnest sense of awe about the brain and its mysteries. The survivor's life emerges as an ongoing pull between the changes that occur within an altered brain and the outward repercussions that follow. It is this tension between being and becoming that begs the intimate, soulful questions posed by every brain injury. What are we other than our brains? Is there a part of me that can't be changed by a brain injury?

My boss, a veteran of the brain injury rehabilitation industry, warned me that most of the people I meet won't be able to find help. Even though it is one of the few brain injury rehabs within hundreds of miles, Brookhaven Hospital has only nineteen rehab beds. Waiting lists are the standard in every brain injury program. If I can't get a patient into Brookhaven, I'm determined to get them somewhere. Somewhere, in my experience, is better than the nowhere they inhabit.

Because of the bureaucratic tangles and the inconsistency of insurance companies, my schedule is a peculiar mix of the arbitrary and urgent. I know that six months from now I will probably be in Dubuque, Iowa, but I don't know if tomorrow morning I will have breakfast at home or next to a truck driver in San Antonio. My schedule is urgent because brain injuries are a matter of life and death. On occasion, the worst happens, and the patient dies before we ever have a chance to meet. Cases close silently, and new ones arrive in their place.

THE HERMIT OF HOLLYWOOD BOULEVARD

When the seizure comes, it hits him like an avalanche, enveloping him in a roaring fog. In a matter of seconds, it sweeps Cheyenne Emerick in its throes and slams him to the ground. The premonition is slight: a tingling sensation, a rushing dizziness, an emotional surge. In the midst of his daily routine, a wave of anxiety and stress overtakes him. Cheyenne's heart begins to thump in his chest and he starts to pace. Even though this has happened a thousand times before, each seizure is more frightening than the last. Cheyenne has survived all the others, but there is no guarantee he'll survive this one.

It could happen anywhere. It has happened in restaurants, in bed, in cars, and in front of dates, friends, and family. Anyone in Cheyenne's presence might bear witness to his most carefully guarded secret, a secret he half keeps in order to protect those around him and half keeps because seizures are the very inversion of physical grace, harmony, and beauty. Sufferers can foam at the mouth, make bestial utterances, and lose their continence. Seizures subject the human body to its most ugly, embarrassing, and chaotic contortions.

"I can only tell how intense my seizures are by the look on other people's faces," says Cheyenne as he wipes his kitchen counter. "They're always, *always* scared out of their minds. I've even seen Dragline shivering in the corner afterwards." Dragline is Cheyenne's English mastiff. On his haunches, he takes up half the small kitchen. It's hard to imagine him frightened of anything.

"Right before I go unconscious, that's when the hallucinations start up," Cheyenne explains. In the early stages of his seizure disorder, Cheyenne didn't hallucinate at all, but now the presence of hallucinations suggests that his seizures have strengthened in their severity. He hears a repetitive thumping sound, similar to the pulsing air from a nearby helicopter, and his mind goes delusional. He may back up to a corner, thinking that he is being surrounded by people who are not there. He may think that he is in another place having a seizure, a delusion of displacement. Recently, he had a seizure in his apartment and thought he was in a faraway city bus surrounded by strangers.

While his mind is elsewhere, Cheyenne's body is completely out of his control. His muscles tighten and his joints wrench themselves so violently that he has torn muscles, sprained ankles, and lacerated his skin.

"I've bitten my tongue so hard that, for several days, the act of eating brought tears to my eyes," Cheyenne says.

The extent of bodily damage depends largely on where Cheyenne is at the moment the seizure hits. If he lands near a wall in his apartment, he might kick through the Sheetrock or lodge his fingers in between a doorjamb. If he only gets a severe rug burn across an ear, it's a stroke of luck. Once he had a seizure while sitting alone in a car and repeatedly bashed his head against the dashboard. He awoke to "a fucked-up face" and a pool of blood in his lap.

At the seizure's most extreme, even breathing is bound and labored. The contortions and the writhing and choking last several minutes, then the seizure fades as quickly and mysteriously as it first arrived. Immediately, Cheyenne starts gasping. The hyperventilating is a gift, a sign telling him that he has survived. After a minute or two, his name comes to him and he begins to piece his reality together again. He becomes aware of his surroundings and of the gaping jaws of bystanders and paramedics. It takes a full fifteen minutes for everything to come back, assuming everything does come back. Cheyenne isn't quite sure what the seizures are costing him neurologically, but he and his mother have a very clear idea of their cost on a personal level.

Most of us have had only a fleeting taste of what a seizure feels like. In the initial stages of sleep, as our mind begins to form hypnagogic phenomena, the fragments of dreams, we creep slowly toward unconsciousness. At some point, our bodies might jerk suddenly as though we've lost our balance. Your arms stiffen, your torso heaves, your legs jump out. The fleeting electrical catch in your frontal lobe that shot your body rigid is the same type of dysregulation felt by people who suffer seizures, only they are suspended in that spasm for minutes at a time, and their "twitches" might occur in any given part of their brain, with hundreds of times the intensity.

Anyone who suffers from repeated seizures, regardless of whether the cause is hereditary or due to injury, has epilepsy, the falling sickness. In 1602, the French physician Jean Taxil made a curious observation. In every case of demonic possession he studied, the possessed individual also turned out to suffer from epilepsy. Taxil did not use the observation to dispute the reality of demons; he merely offered the insight in hopes that others might notice the devil's penchant for epileptics. We have since learned that God and his angels also consider epileptics a hot property.

"Something divine can be observed in epileptics," wrote the surgeon Fabricius Hildanus,[1] "for very often, something lies hidden in the bodies of epileptics which is above our power of comprehension." Sacred literature is replete with prophesies, mystical texts, and holy visions that were contributed by such assumed epileptics as Mohammed, St. Paul, St. Birgitta, Mechthild of Magdeburg, Hildegard von Bingen, and St. John of the Cross. It was William James, a prophet of a different sort, who most eloquently defended the epileptic's supernatural states in his seminal work, *The Varieties of Religious Experience*:

> To plead the organic causation of a religious state of mind, then, in refutation of its claim to possess spiritual value, is quite illogical and arbitrary . . . Otherwise none of our thoughts and feelings, not even our scientific doctrines, not even our *dis*-beliefs could retain any value as revelations of the truth, for every one of them flows from their possessor's body at the time.[2]

James goes on to suggest that atypical physiological states might actually operate as gateways to unique, legitimate insights and that to conclude otherwise is obscenely reductive.

"If there were such a thing as inspiration from a higher realm," he says, "it might well be that the neurotic temperament would furnish the chief condition of the requisite receptivity."

For years, Cheyenne's ictal experiences gravitated more toward the demonic than the heavenly. Where others sense a cosmic oneness, Cheyenne gets gypped with conflict and despair. Some epileptics achieve a nearly orgasmic euphoria; Cheyenne gets a migraine. He thinks that the dark visits are a type of karmic retribution, a payback for his proclivity toward self-destruction, as though a seizure opens up a personal Pandora's box of pent-up emotions. While few neuropsychologists would side with him, Cheyenne blames a misguided childhood, followed by a reckless adolescence, for his more nightmarish seizures. Only recently have Cheyenne's seizures undergone a transformation.

The lush palm trees covering Cheyenne's place are misleading. His apartment is a shithole, a tiny flat lodged atop a dilapidated house not three hundred paces from Hollywood Boulevard. The thin, chipping planks leading up to his door groan with each step I take, and the swaying handrail only compounds the guess that I might drop through the staircase at any moment. At the top of the flight, the metal screen door is propped open, and Dragline stands in the doorway and affectionately swathes my corduroy pants with a brown, grainy coat of drool when I call Cheyenne's name.

"Be right there," he calls from around the corner.

In a long-sleeved white shirt and blue jeans, Cheyenne looks unassuming and kind. His features are soft and his frame is still boyishly thin, an echo of his past as an athlete.

I don't see much California in him, not with his pale skin and unkempt thatch of blond locks, but I do notice the multiple nicks and healed-over scratches that pepper his neck and face and hands. He looks like a man used to being roughed up, which makes it easier for me to see the playwright in him. Without looking up at me, Cheyenne grabs Dragline by the collar and pulls him aside to make room for me to walk in. After a few mumbled apologies about my pants, he hands me a dish towel to soak up the dog slobber and begs me a few steps farther inside.

Cheyenne lives minimally, but not by choice. The few items in his apartment are either donated or found objects: a small television, a twin mattress, a Mr. Coffee coffeemaker, a metal oscillating fan, and a few dumbbells. A relic from former days, an abused snowboard, leans in the corner. Cheyenne feels lucky to have a landline telephone; they didn't check his credit when he called for service. Six steps into his doorway, and you can count the sum of Cheyenne's possessions.

In such small quarters, Cheyenne appears tall enough to divide a room, but his unobtrusive manner retracts the space around him. It's a practiced effect that many brain injury survivors achieve, sometimes to the point of invisibility. Cheyenne, however, has a restless quality that's hard not to notice. After he offers me a seat, he flips open his cupboard and asks me whether I want coffee or water, and although I rarely drink coffee, I ask for it so he'll have something to do. It gives me a chance to watch his motor skills and check out his initiation and sequencing. It takes a certain cognitive level to make coffee, and Cheyenne does it like you would do it, only when Cheyenne does it, it's a relief to me. I like to see brains working well, especially when they shouldn't.

"I've been tired all day, so I just had some iced tea to help wake me up a little," he explains. His hands have a caffeine tremor, as opposed to a marked palsy. He continues talking to me as he makes the coffee, and the complex tasking is another sign. Already, he's in the upper fifth percentile of people I evaluate. It's a pleasure to see him moving and acting so fluidly. Clinically, he presents well, which is another way of saying that it's my guess he's hiding a lot. The only way I'm going to get a good read on Cheyenne is by letting him do all the talking, but I already know the end of our conversation is going to consist of apologies and downcast eyes.

We sit down together at his kitchen table, and Dragline parks himself in front of me and looks at me eye to eye. This is Cheyenne's confidant, I think. He sees the real Cheyenne, day in and out, and he's not afraid. Like so many other people with brain injuries, rage is a part of Cheyenne's injury, but I don't see any signs of it when Cheyenne reaches out to pet him. Dragline leans a little into Cheyenne's hand, and Cheyenne is surprisingly gentle and affectionate with him. I take a cue from Dragline and lean back in the chair, hold the coffee mug in my lap, and cross my legs in a therapist's pose.

"So what's snowboarding like?" I ask him.

Sugarloaf Mountain in Carrabassett Valley, Maine, is a breeding ground for world-class skiers and snowboarders; it's one of the few mountains in New England that garner respect among West Coast aficionados. Known for its slick, icy runs and its steep vertical drops, Sugarloaf lures snowboarding extremists hell-bent on speed and more speed. As early as the late eighties, Cheyenne was regularly hitting sixty miles per hour on the mountain's trails.

"Back when I started, snowboarding was still in its infancy," says Cheyenne. "People used to stop me and ask what I had strapped to my feet."

He earned his chops carving the slopes of Sugarloaf in the late eighties, but as a sport, snowboarding had few real fans and even less respect. Cheyenne and his pack dismissed local competitions as pointless and demeaning, preferring instead to build their own jumps and improvise their own runs along the mountain's uncharted side.

"Because we didn't compete, we had no idea what kind of skill level we were riding at," Cheyenne says, "until we got a visit from *Boardrider*."

The *Boardrider* writers were taken aback at the sight of Cheyenne and his friends ripping across ice-packed terrain in their baggy T-shirts and ski goggles (at the time, helmets were considered laughable). They wrote a few sentences about the pack of renegades, commenting on their penchant for risk taking. Cheyenne and his buddies joked about the modest mention they had in the magazine, but soon they found themselves fitted in the newest gear from a sponsor, even though they maintained their anticompetitive stance. As the buzz around them grew, they sensed the time was drawing near. For years, they had set their sights in the same direction: Snowbird. It was the most daring, dramatic mountain the West had to offer, and it was a magnet for the most magnificent snow in the states. When other surrounding mountains got inches, Snowbird always got feet of dry, brilliant powder that padded the entire slope. After the hard-packed slams Sugarloaf dealt him, Cheyenne reckoned that it would be impossible to get hurt anywhere on Snowbird.

. . .

Flatlight is a type of blindness feared by mountaineers and skiers alike. The high, gray-white sky fuses with the icy terrain, and from certain angles, ground and sky cancel each other out. In flatlight, hilltops vanish, crests and valleys disappear, and the ground ahead looks like an incandescent blur. All sense of depth perception is lost. Hit the slopes at the wrong hour of the day, and you may not see any moguls, tracks, or other telltale signs of the uneven terrain ahead. Because flatlight is a standard hazard for New England riders, Cheyenne and his friends had become overly familiar with the phenomenon. Flatlight never slowed their runs back home, and it certainly wasn't going to interrupt Cheyenne's first run at Snowbird Mountain.

The approach to Snowbird was a sacred moment for Cheyenne. He had already singled out all the right gear the night before. He woke to an alarm clock, the first time in years, and peeked outside. The red morning sun was receding into the cloud cover, hinting at the gray day to come. Cheyenne stretched his legs back while his friends dozed on couches and sleeping bags. He showered, collected his gear, and crept out the door without disturbing anyone. Outside, he paused to listen to the avalanche cannons booming across the valley.

As he rode up the lift, Cheyenne studied the descent, just like he had done a million times back home. He noted that he might want to veer left at that tree with the broken branch, so that he could clear the small patch of boulders below. The drop just beyond the tree looked like a creamy spread, and the landing below like a pillow. From the lift, the jump seemed like a lark—twenty, thirty feet at best. It had snowed several feet the night before, so the landing would be soft and forgiving. The rest of the slope opened up like a six-lane stretch of highway, perfect for bombing.

With a small hop from the lift and a few digs past the skiers, Cheyenne locked his other foot into the binding and eased onto the slope. His board sank deep into the loose powder and the feeling of snow up to his waist was heavenly, like wading through feathers. As he picked up speed, the acceleration levitated Cheyenne up toward the surface of the snow. Until that moment, he had only heard what snowboarding in deep powder felt like. Now, on the powder's edge, Cheyenne became the focal point of balance and grace. The snow billowed out behind him like a jetstream as his body swayed to the contours of the mountainside.

On the horizon, he spied the tree with the broken branch. He was approaching it so fast he hardly had time to remember to adjust his path ten or so feet to the left, to hit the drop-off just right. He tucked and shotgunned it toward the pine, cranked his board ever so slightly to the right, and shot off the cliff in a burst of snow and ice.

If riding at top speed elicited singularity, then flight is the vanishing point. Here, soaring above the treetops, was absolute liberation. There was no thought, no worldly connection at all. In that moment, he was the mountain, the sky, the snow, light and lightness, all of it happening at once.

"When you're going thirty to forty miles an hour and you fly over a cliff and you lightly touch down on a bed of snow, it's really an incredible sensation," says Cheyenne. "There's no other feeling like it on earth."

He cleared the jump effortlessly and watched as the run widened before him. The slope ahead was clear and unobstructed by skiers—it demanded full speed. Cheyenne tucked down on his board, aimed his left hand directly above its nose, and positioned his body for minimum wind resistance. He rocketed down the slope, picking up speed as the ground rushed below him.

At peak acceleration, Cheyenne's legs compressed lightly, giving him only a split second to understand that he had shot another jump, one that he overlooked on his ride up. He crunched over on instinct and launched into the unexpected air.

After what seemed like the right amount of time, Cheyenne's first thought surfaced.

Where is the ground?

Cheyenne tucked his knees up and pulled the nose of his board back with his fingers. He peered over the edge and saw only a bright gray field.

Where the fuck is the ground?

There was no telling where the sky stopped and the ground started. He glanced to his side hoping for a hint from the trees, but they only compounded his disorientation as they whipped past him in an evergreen blur.

Cheyenne was still looking down in a hunched-over, quasi-cannonball position when the ground rushed up from nowhere. The force of the impact drove his knee into his forehead, blasting away all his senses in a burst of white pain. Cheyenne's brain slammed into the inner front wall of his skull, against a particularly coarse, ridged patch of bone. His right prefrontal cortex took the bulk of the impact, while his entire brain rebounded against the back of his skull.

Cheyenne hit the ground so hard that his body completely submerged below the powderline, then ricocheted back out, sans board and bindings. His body came to a dead stop, but the momentum yanked both his gloves off his hands and sent a boot flying fifteen yards away. In snowboarding parlance, Cheyenne had a yardsale—a crash so intense that personal effects are strewn over a wide area. In a grimmer manner of speaking, Cheyenne had just suffered a catastrophic injury.

The impact left him dazed and dumbfounded, but somehow his body collected itself along with its riding paraphernalia and wobbled down the rest of the mountain, just as automatically as he had taken any other fall back home in Maine. When he came to his senses again, Cheyenne found himself on the lift, preparing for another run. Typically, he left the guardrail up on his chair so he could adjust his bindings, but halfway up this ride, Cheyenne felt a rush of vertigo. He reached up to the rail in midswoon and pulled it down in front of him.

"I will never, ever forget the feeling of my first seizure," says Cheyenne as he leans back in his kitchen chair. He moves an uneasy hand through his hair and then rubs the back of his neck. "It was a feeling of utter, total, impending doom. I thought I was losing my mind."

That day, Cheyenne suffered several more seizures as he continued to ride down the mountain. Each one manifested as an intense, dizzying spell of anxiety, paranoia, stress, and fear. He told himself that he shouldn't have partied so hard the night before, that perhaps his fast and hard ways were catching up to him. The more runs Cheyenne rode, the further his crash landing slipped from his memory, until he had forgotten it altogether. Even while talking to his mother that evening, he neglected to bring it up. But he did tell her about the strange spells he had.

"I have those all the time. It's just anxiety," Holly Emerick explained to him. "I'll send you some homeopathic pills that I take. They'll really help."

The explanation was good enough for Cheyenne. In his mind, the emotional meltdowns became merely a casualty of aging. He took the homeopathics along with his beer in the evenings, but the seizures still came steadily, averaging five or six times a day. Cheyenne taught himself to remain composed when he felt the onslaught. His friends just thought he was a little distracted, or maybe he was drinking a little more than he let on.

Cheyenne carried on the charade, continuing to ride out the season, then the summer, and another full season.

"I had attacks while I was riding," Cheyenne says. "And I just kept riding through them. Each time a seizure happens, there's a lingering effect of stress in between the attacks, because you don't know why they're happening to you. And they continue to happen. It's very disturbing. You don't just go back to normal. It leaves you unsettled, it leaves a mark on you."

Seizures occur on a continuum of sorts, but their murky organization is a hint at how little we know about and understand them, especially in relation to consciousness. First, seizures are divided into two types: epileptic seizures and psychogenic nonepileptic seizures.[3] Of the epileptic variety, seizures branch off into primary generalized seizures and partial seizures. During a primary generalized seizure, electrical discharges occur throughout the brain, unlike partial seizures, which concentrate in one lobe. Partial seizures are typically the result of trauma to the brain, while primary generalized seizures are often hereditary, though they can also be the result of injury.

There are three types of partial seizures: simple, complex, and secondarily generalized. Short-lasting and relatively common, simple partial seizures occur while a person remains conscious of their surroundings and of their seizure experience, and may include auras, smell hallucinations, emotional turmoil, or peculiar states of mind, such as profound déjà vu. Complex partial seizures differ in the sense that a person's awareness of the event often evaporates, so that they appear to onlookers as though they're caught in a daydream. Sometimes, complex partial seizures will induce individuals to engage in bizarre, automatic motions like lip smacking or head shaking—one unfortunate patient I met unconsciously disrobed during his seizures. Secondarily generalized seizures, the third type, are essentially partial seizures that graduate to a more severe kind of seizure.

Among the seven kinds of primary generalized seizures is the petit mal seizure, or absence seizure, which is a short, forgotten moment of staring—most people are completely unaware they've had a petit mal. When a seizure involves only muscular stiffening, it's called a tonic seizure, but when the tonic seizure is followed by muscle contractions, it becomes the most renowned and feared seizure: the grand mal, or tonic-clonic seizure. During a grand mal, an individual is unconscious while the body racks itself in a joint-torquing frenzy. The only thing worse than a grand mal is a grand mal that won't quit.

Convulsive status epilepticus hangs over Cheyenne's head like the sword of Damocles. Textbooks regularly state that seizures over five minutes in duration require immediate medical attention. Unchecked, a *status epilepticus* is a horrific seizure loop that won't end after five, ten, fifteen, or even thirty minutes. During *status epilepticus*, a grand mal seizure builds and builds and doesn't subside until its host is either dead or hospitalized. It is a life-threatening condition that locks onto the brain and denies it any escape. Should a person survive *status epilepticus*, they stand a grave chance of severe brain damage, including persistent vegetative state.

In this classification scheme, Cheyenne suffers seizures from both sides, partial and generalized. Following his accident, he endured both petit mal and simple partial seizures several times a day. While he continued to snowboard, his seizures gradually ate away at his well-being, always undermining his ability to fully enjoy his life for fear of having another "panic attack." He never brought the attacks up, even in private conversation, and refused to let them interfere with his lifestyle. Denial worked beautifully for a full eighteen months, and then it stopped working altogether.

Cheyenne is standing on the front porch with a beer in his hand. The sunlight feels good against the back of his neck, and he and his friends are making plans for the trip back home. Cheyenne lifts the can of beer to his lips, and this time the quickening is intense and doesn't let up or even plateau. A tidal rush of electricity burns from his head and there is no containing this seizure, not this time. His universe goes white.

When he comes to, he's in a puddle of blood and beer. His forehead is throbbing and his face is pressed up against the hot pavement. A man is talking to him and feeling his neck but he can't make out the words. The man pushes a wad of gauze and cotton onto his cheek and it turns bright red in seconds. Cheyenne can't seem to figure out anything, none of the questions the man asks make any sense. The flashing lights and the fear-stricken faces follow him as he is lifted first onto a gurney, then into the ambulance.

By the time he gets to the hospital, the blown-apart pieces of his life are drifting back together. His name is Cheyenne.

This is Arizona. It is June. He was at a friend's house. He fell on the ground. He must not fall asleep. Yes, he can hear you. Yes, it hurts. No, he doesn't have any medical conditions. No, not diabetic. No family history. Only two beers. Maybe three. Nothing compared to normal. No, just panic attacks every once in a while. What? No, not married. No kids.

At the emergency room, the routine is all business. The EMT briefs the trauma nurse in whispers, then passes her a sheet with vital signs. She snaps on latex gloves and marches straight to Cheyenne and rips off the gauze and the corners of her mouth go down only a little. She preps the wound with a splash of hydrogen peroxide and dabs alcohol wipes around his face. Cheyenne doesn't flinch at the sting; he hardly notices anything past the pounding in his temples. A resident shows up, and zip, zip, zip, the gash is closed and dressed and Cheyenne is ushered off for labwork.

First he's hooked up to an electroencephalogram, and after a few minutes worth of hieroglyphic spikes, they schedule him for his MRI. It's just hospital policy after any fall, after any bumped head. It doesn't mean anything, they tell him. Better safe than sorry. The neurologist will talk to him tomorrow about the results, another formality. Cheyenne turns down a prescription for a painkiller, and he shakes hands with the doctor and walks out the door of the hospital with three stitches on his face and a mean headache.

Within twenty-four hours, Cheyenne is back at the hospital for his MRI. He empties all his belongings into a small tray, gets rid of every hint of metal attached to his body, puts on a flimsy gown, and inserts foam earplugs. He's escorted into a room specially designed to withstand powerful electromagnetic fields, and steps up onto a table. The technician turns Cheyenne's head slowly and carefully, like it contains a warhead, then leaves the room, and fires up the magnet. As the magnet whirls around Cheyenne's head, it sounds like a freight train, even with the earplugs. It doesn't feel like anything at all, and when it's all over, Cheyenne is back in his clothes with an appointment card for the neurologist in his hand.

The neurologist's office is small and impersonal and full of old people. He waits and waits and is finally placed in a windowless room and plays with the plastic model of the brain on the counter. The neurologist comes in and doesn't raise his eyes. He has an envelope full of films that he begins to thumb through. Cheyenne is uninterested but figures he could use a doctor's take on his panic attacks while he's here. The doctor sets down the films, glances through the MRI summary, and finally speaks.

"You have a scar on your brain," he says, finally looking up at Cheyenne as though he suddenly appeared. "On the right side of your frontal lobe, and it looks like it stretches down into the temporal area. The other day, you had a grand mal seizure, and that's most likely where it's coming from. It's a permanent condition called traumatic brain injury."

"I was in total shock," Cheyenne recalls. "For a moment, I just sat there stunned, and then it sank in."

Cheyenne had been in the hospital multiple times over snowboarding-related injuries in the past few years. He had cracked vertebrae, broken ribs, and had even compound-fractured his wrist so badly that bone pushed through his skin. He had endured a host of horrific injuries with only an occasional grimace or wince, but when he heard the news about his brain, he folded his hands around his face and cried.

The doctor rolls his eyes and looks at Cheyenne as though he were something that had just been scraped off his shoe. He mumbles a few things about putting him on an anticonvulsant, which may or may not help, who knows. Then the doctor shoves the films under his arm and closes the door behind him. Cheyenne stays in the office, crying and confused by the doctor's disregard, until a nurse knocks on his door and asks him to leave.

"I was hysterical," admits Holly, Cheyenne's mother. "The doctor told me Cheyenne had frontal lobe brain damage that was irreversible and that he would have epilepsy his entire life. I thought I was going to faint. I got sick to my stomach, and I started crying and crying. It was a nightmare I couldn't wake up from."

Holly rents a room in a small home on a wide-open range in northern Arizona. A colossal, snow-capped mountain splits the view from the dining room window, but Holly acts as though she doesn't even notice it, it's so familiar. With long graying hair and a face unmarred by makeup, she has the hardy poise of a life lived outdoors, but her expression bears the heavy lines of perpetual and serious worry.

"Look how hard it is for people in this country, then you add a head injury, a disability? I didn't want that for my kid," she says. "Since then, I haven't had a real day of rest. How can anyone rest when their child has a life-threatening condition?"

I nod my head and return to the spinach quiche she served me. I don't have an answer for her. Medicaid has denied them, and Cheyenne's injury is too invisible for disability income. Holly is so overextended she no longer qualifies for a bank loan, much less a credit card. She is in her sixties, she works two jobs, she's never going to see retirement, and

I'm eating her quiche and drinking her tea and nodding my head. I'm just here to listen, I remind myself.

After the doctor's consultation, Cheyenne's family flew him back to Maine, where he joined them. Although superhuman resolve had been pounded into them over multiple Maine winters, Cheyenne's condition tested their endurance.

"We could tell things weren't the same," says Holly. "The new medication was causing him to stutter. He struggled so hard to get words out that we put pencils and paper all over the house for him to use."

A seizure has a way of entangling everyone in the vicinity. The epileptic is caught in the bodily spasms, but everyone outside the seizure is held spellbound by the responsibility to do something, anything, even though nothing can be done to assuage the intensity or duration. You can roll a person onto their side, and you can move objects out of their way.[4] Although they may bite their tongue, they won't swallow it—anything placed in the mouth of a seizing person is liable to cut or choke them, not help them. Outside of damage control and keeping a distance, the only other option an outsider has is to wait. Cheyenne's grand mal seizures last from four to six minutes, and they are some of the longest four to six minutes in any observer's life.

"The first time I saw him have a grand mal, I knew what was happening and still I started hyperventilating," says Holly. "We were in his father's apartment, in upstate New York, and he was in another bed in the same room. He was talking to me, and his voice started getting louder and louder, and he was talking faster and faster, and then he stopped."

From the dark corner of the room came high-pitched whimpering noises, like a trapped animal's whine. Then came low, growling noises and gasps and sputters. Holly turned on the lamp and saw her son's eyes flutter and then lock open, the beautiful blues replaced by milky white. His jaw was distended in an agonizing turn and his hands were flapping against the sheets. She grabbed a shoulder and rolled him onto his left side. Foamy saliva poured out of his mouth, clearing his airways.

She dialed 911 and paced nervously around the bedroom, keeping a short distance between herself and her son. Their dog, a golden retriever, ran in circles in the other room. The emergency crew arrived quickly, while Cheyenne was still seizing, and even they were brought to a standstill. They stood and watched, mouths open, until the convulsions waned into shudders and Cheyenne lay limp on the bed. Color returned to his cheeks and the blue irises returned.

The frequency of most seizures can be reduced with a variety of antiepileptic medications, but each of them comes with a list of undesirable side effects ranging from blood disorders to permanent involuntary spasms. Cheyenne felt that phenytoin offered the best compromise for his seizures, but the effectiveness of phenytoin depends on keeping a constant level of the drug in your system at all times. At the hospital, Holly learned that Cheyenne's phenytoin levels had dropped to a dangerous low, and that he had to remain in the hospital until his levels were brought back up through a slow course of intravenous injections.

"After you witness something like that, you feel like your life is to some extent out of control," says Holly. "And the scary reality is that it's true. The welfare of your child is out of control. It's very hard to be away from him, and that's part of why he agreed to go to school."

It was easy to blend in at Arizona State University. With a student body the size of a small town, it was a place where

Cheyenne could easily hide his brain injury. He didn't check in with ASU's disability services coordinator, he didn't inform the school that he had a seizure disorder, and he didn't talk about his injury to professors or peers. Even after his grand mal seizure had outed him as an epileptic among his friends, Cheyenne's new environment allowed him to once again pretend his disorder didn't exist. It was a great act, one that fit naturally with Cheyenne's other knack: theater.

Cheyenne hit the drama department with the same aggression as he hit the slopes back home. He honed his skills backstage through semester after semester of production involvement, but diligently shied away from actually performing in front of people until another student cornered him for a short, offensive one-act play she had written for her senior project. The prospect for Cheyenne was terrifying and exhilarating. None of the department teachers had ever seen him act, and none of them had ever seen him have a seizure. With a mouthful of caveats, he accepted the role and showed up every day for rehearsal. He told himself that if he felt a seizure coming on, he would just quietly walk offstage and seize, end of show.

The seizures didn't go away, but they did seem to make concessions. During rehearsals, Cheyenne displaced his brain injury so thoroughly that he never had an ictal event, not even the telltale tingling. Onstage, he became his character, and his character didn't have a seizure disorder. Offstage, however, was a different matter. The partial seizures haunted him in between classes, through back hallways and in the corners of locker rooms.

The grand mal outside of the weight room nearly killed him. He had been working out in the school gymnasium, with extra weight and extra reps. Snowboarding had taught him how far he could take his body and how great it felt to push its limits. He worked himself hard that day, and during the second rep of the second set on bench, doing butterflies, the light shifted and his breath caught. He quickly set down the weights, picked up his towel with a trembling hand, and marched out of the room and down the long linoleum hallway in hopes of making it to the locker room where he hoped to have his seizure in private. The last thing he wanted was to seize in the weight room and then get banned from working out altogether.

He blacked out before he made it to the locker room. The momentum of his stride in tandem with the rapid onset of his seizure launched him midpace onto the linoleum headfirst. The convulsions kicked in and slapped his face repetitively against the ground. Cheyenne's hard-earned strength worked against him now. His back and neck muscles tightened and retracted over and over again and he yelped and groaned. The commotion sparked a phone call and soon a crowd. When Cheyenne finally came to, paramedics and nearly every person in the gymnasium surrounded him. The pain that typically radiated from the inside of his head now blended with pain on the outside of his head. It hurt to open his eyes. He let the paramedics transport him to the nearby ER to get his levels checked and his swollen face and lips ice-packed. During the ride there he thought about the next time he would exercise, and how everyone would avoid looking at him and would keep their distance.

Over the weekend the swelling subsided enough for him to claim allergies and a lack of sleep when rehearsals resumed. He counted on the performance as much as it relied on him. Play practice became a refuge, an arena that allowed Cheyenne to tap into the wilder currents of his emotional life. In the quiet darkness of a theater, he could prove himself to a public that dismissed him as damaged in the daylight. On the day of the actual performance, the stage lights narrowed and Cheyenne found himself surrounded by complete darkness on all sides. He could hear the audience's coughs and shuffles, but could only make out shadows at best. He knew that if the lights were to flicker or if his blood pressure were to rise, it could set off a seizure before he even began his first line. Just thinking about it—or not thinking about it—might cause it.

Cheyenne's presence filled the stage and captured the audience of students and professors. He ranted through obscene passages, howled his own angst into his character, and spent his last moments panting and fuming postmodern pensées. When he delivered his last sentence, the lights lingered over his collapsed frame and slowly faded. The darkness held constant and quiet. Cheyenne stood up to brush himself off and was nearly bowled over by the thunder of applause. The lights lifted to reveal Cheyenne's first standing ovation, an ovation he answered with a slight bow and his own applause to the producer.

Cheyenne retained his major but eventually transferred to the University of Arizona for a brief time, where one of his teachers immediately picked up on his raw talent, calling him a young Sam Shepard, a new David Mamet. The professor pulled him aside one day and whispered that he ought to consider bypassing school and heading straight to New York or Hollywood. The prospect struck him as ludicrous, impossible, unwise, and extreme. Cheyenne had to do it.

In the crammed hills of Hollywood, Cheyenne found himself utterly alone. He had arrived with barely enough money in his bank account to make a down payment on rent, but he planned to get by on night jobs so he could audition during the day. Tens of thousands of people in the valley were grinding at the same wheel, and he planned on grinding harder than any of them. Cheyenne's spirit was firm and resolved, but his brain was ill-prepared for the challenge. Seizures aren't the only problems that result from a banged-up prefrontal cortex.

Frontal lobes act as filters, and Cheyenne's filter has a big tear in it. When undamaged frontal lobes encounter windshield wiper noise or restaurant chatter, they throw up a neural muffler that allows a person to tune out unimportant stimulation. Over time, even the most unobtrusive sounds can provoke intensely agitated states in TBI survivors. The filter goes both ways, too. People with frontal lobe injuries regularly report that they're unable to stop making statements that they'd rather keep quiet. One patient I evaluated told me that living with a frontal lobe injury was like living with a foot in her mouth. Cheyenne might just as easily tell one of his diners to have a nice day or to shove his menu up his ass.

At school, Cheyenne had the advantage of controlling his environment: he set his own schedule, he studied at his own pace, and he participated only in activities that made him comfortable. Student life rarely tested the full capacity of his neuropsychological functioning, whereas even the most menial jobs in Hollywood required some level of cognitive fluidity, a level that Cheyenne found unattainable. As a waiter, he couldn't jot down orders correctly because of all the background noise. If a partial seizure occurred during work, he might forget to flip over a steak or lock up the front door. His bosses said he was too absentminded, too rough with customers, too distracted. They told him he didn't fit, they reduced his schedule to a pittance, and then the restaurant manager suggested the construction site down the road. A week later, the foreman there eventually directed him to a moving company. Over the course of years, his résumé read less like a career portfolio and more like a guide to grunt-level jobs in Los Angeles.

It took money to audition, and when the choice came as to whether he should eat, buy his medications, or pay for headshots, Cheyenne chose to eat. The less medication he took, the more frequent his seizure activity became, causing him to lose the work, money, and benefits he needed for medication. Cheyenne reconciled himself to the idea that he would settle just for surviving in Hollywood, but deep down, he wasn't sure he was capable of even that.

One of the world's most distinguished epilepsy treatment programs in the country is located within a short walking distance of Cheyenne's apartment. Cedars-Sinai Medical Center is a monolith that juts out of the valley skyline and is visible from most parts of the city. After he first moved to California, both Cheyenne and his mother began calling Cedars-Sinai in hopes that Cheyenne might be able to join their program.

The Epilepsy Consultatory Specialty Clinic at Cedars-Sinai is a patient's wish list fulfilled. The treatment team consists of epileptologists, neuropsychologists, neurologists, and an army of clinicians. At their disposal is an arsenal of diagnostic medical equipment that can detect, scan, image, probe, and trace every measurable aspect of a seizure. The clinic is regarded as one of the foremost practitioner centers for Ictal SPECT, a procedure wherein a radioactive element is injected into the bloodstream so that seizure activity lights a video screen like a Day-Glo weather map. By ascertaining the exact location, duration, and intensity of seizures, neurologists can forgo the common trial-and-error route and instead offer specific treatment strategies customized to each individual patient.

It is the holistic ideology of the program, and not the equipment, that makes Cedars-Sinai tower above typical treatment centers. The goal for every patient is to become seizure-free, so ongoing treatment targets events that may set off seizure activity. A person's lifestyle is scrutinized for triggers. The culprit could be a monitor on a home computer that flickers at a rate that irritates a sensitive area of the brain, or it might be a certain beat on any given CD that primes a person for a seizure. Treatment consists not only of avoidance strategies but also gut-level lifestyle changes. Multitasking may no longer be an option, certain jobs or careers might be off limits, and some clear boundaries may need to be drawn in particularly incendiary relationships. Stressful events potentiate ictal activity, turning some seizure sufferers into suburban relaxation gurus.

Save for his financial standing, Cheyenne is an ideal candidate for treatment at Cedars-Sinai. An initial consultation without health insurance runs close to five hundred dollars, and the consultation would simply validate his need for additional testing. Labwork, which would include a complete battery of imaging, would cost him several thousand dollars. Add to that the ongoing cost of medication stabilization, routine blood monitoring, and occasional checkups, and Cheyenne's treatment could run into tens of thousands of dollars, dollars that no family member or friend has at their disposal. Given that even the local cable company refuses to offer him service, he isn't likely to find any consolation from the CFO at Cedars-Sinai. To Cheyenne, Cedars-Sinai exists only as a visual taunt, a reminder of his life of constraint.

A few times a year, Holly places a phone call to Cedars-Sinai in hopes that something may have changed, that there is an affordable new option. Holly knows what she will hear, but a phone call doesn't hurt and the number is toll-free. I asked her what Cheyenne's injury has cost her in actual, earned dollars. She estimated that it has personally cost her more than fifty thousand dollars so far, not counting the tens of thousands of dollars in credit card debt that hound her. It has cost her ex-husband, Cheyenne's father, an equal amount of money, so much that although he was a successful business owner at the time of Cheyenne's injury, today he can no longer afford basic health insurance for himself. Several friends of the family have assumed debts ranging from several hundred dollars to upwards of thirty thousand dollars and counting. They have received help from everyone they know, and yet Cheyenne opens the door to an empty fridge on occasion and hopes that tomorrow might bring a quick-paying job.

Holly doesn't need to mention the opportunity costs of the injury; it is all right there in front of me. The last-stage car, the borrowed furniture, the angry expression she gets when she talks about her experiences at the local Department of Health and Human Services office.

"You feel like you are living in a dark basement. We struggle day to day, we pray that he stays alive, that we can earn enough to get the medical care he needs," says Holly. Her jaw is clenched and she's looking out the window, not seeing the mountain again. She places her napkin on the table and smoothes it out in slow, comforting strokes.

"I have noticeably aged in the past few years," she says. "The bills are taking a terrible toll on me emotionally, and I am stretched to the limit. I know that his injury has brought us together, but trust me, there are better ways to deepen your bond with your children."

Starvation, isolation, depression, anger, existential angst, mortal dread, and dismal luck with women . . . Cheyenne had all the prerequisites needed to become a playwright. In the long droughts between jobs, he turned to his prehistoric computer with the eight-inch monitor and started exorcising his despair into words, and the words came out as a conflict-ridden dialogue. Over the course of months and months, the dialogue summoned characters bedeviled with anger and hatred. The characters chewed at each other in the first act, spit out their innards in the second, and set fire to their own hearts in the third. *The Death of Don Juan* won Cheyenne his first produced play in Hollywood, and while it brought in sellout crowds, the valley swallowed it whole, repaying Cheyenne with the pittance of a writing credit, a few ovations, and little else. I read the entire work on the flight west and barely move a muscle. Reading Cheyenne's play is like watching the most vile and unlovable parts of yourself spread out for the world to see. It's the closest I've ever come to knowing what his seizures must feel like.

By the time we meet, Cheyenne has been living in Hollywood for more than four years, and his seizure activity has increased to a hundred and twenty events a month or more. Acquaintances from past jobs occasionally drop by to collect old debts, but end up settling for Cheyenne's stereo or baseball card collection. In darker moments of jobless, foodless despair, he has taken a hammer to several of his few remaining possessions. More than once, he has thought about walking straight into the emergency room at Cedars-Sinai and committing himself to the only floor that would take him free of charge, the psychiatric crisis unit. Instead, he paces around his apartment, his hands alternately clawing between his head and heart. In his darkest states, a seizure often takes hold of him, and in its provisional escape, that seizure becomes a mercy.

"There are days I wake up and wonder who I'm kidding," Cheyenne says. "There are thousands of people out here better looking and more talented than me, and they're taking things for granted. They can wake up in the morning and go have a cup of coffee with someone, but I can't even land a date. My poverty and my anxiety have made me very insecure. A girl who doesn't know about my past can still sense it when she is around me. I get tired of having to explain myself, trying to make up reasons and excuses why I don't have furniture, why we can't go out to the movies. No girl wants to lie on the floor and watch the fan move back and forth."

Prior to my own visit, the picture both Cheyenne and his mother painted for me was grim. From a previous job on a psychiatric ward, I've internalized a suicide risk checklist that rings bells every time I hear graycast words. On paper, Cheyenne met all the criteria for suicidality, but when I talk with him, I don't sense that terrible vacuum of transference that envelopes the truly suicidal. I ask him, in all earnestness, why he seems okay today.

"Well, it's strange you ask, because I do feel different," he says. "Something has definitely changed recently, and I can tell you the exact moment it happened."

The week before I arrived, Cheyenne had taken a job making local deliveries for a small janitorial supply company. One of the stops along his route was Nightingale Cottage, a nursing home and hospice for end-stage seniors. He loaded up a dolly full of napkins and Styrofoam cups and buzzed the front door to gain entry. The lock clicked and he pushed on the door, going back-first from the bright sunlight into the dim gloom of the lobby. The smell of urine and open wounds made him hold his breath as he wheeled the dolly toward the kitchen. When he turned a corner, he saw a foyer of elderly people corralled in a circle, none of them appearing conscious of one another, nor of their surroundings.

"They were just staring into space, waiting to die," Cheyenne explains. "And it hit me right between the eyes, it hit me so hard that I just stopped and stared. That was me. In that wheelchair, that was me. It was so clear and so horrifying. That was my future, that's where I was headed."

The Christian mystic Dionysius the Areopagite wrote: "Entering the darkness that surpasses understanding, we shall find ourselves brought, not just to brevity of speech, but to perfect silence and unknowing." If Cheyenne's moment in the nursing home was an entry into the same darkness, then perhaps he stood a chance. A chance for what, I don't know, but it's a chance that he didn't have before.

As Cheyenne tries to explain what happened in that instant, I scrutinize his face. His eyebrows rise a little, his jaw becomes slack. He really didn't know what hit him, but he could still feel the difference in his mind, the unknowing. He tells me everything is different now, that the universe is on his side, and I nod my head and smile. Cheyenne is no mystic. His voice is so full of conviction that I cannot get myself to believe him. I stopped trusting conviction years ago.

"I had a different kind of seizure the other day," he tells me. "It left me in awe, it was so singular. It started going off with such a unique intensity, it was the intermingling of joy at the same time as doom, it was beyond description. Then I saw myself as though I were someone else, and it was—"

Cheyenne finishes his statement with a shake of his head. He's been rendered silent. Perfectly silent.

After our conversation, Cheyenne and I walk down Hollywood Boulevard, looking for a quick bite to eat before I head back. The recent showers cleared out the air and washed the sidewalks so that the Walk of Fame stars look pretty for once. I feel odd stepping on Susan Sarandon, like she doesn't deserve it. We talk a little bit about writing, and about how you can't trust your convictions, and it dawns on me that Cheyenne doesn't trust himself either. He doesn't know whether things have really changed for him or not.

"I guess we'll find out in a few weeks if this change sticks," he tells me.

"It might," I reply, "but I'll be checking up on you. I want to know, for myself."

We each order a slice of pizza street-side and start back toward his place. I listen to Cheyenne talk about the current play he's working on, how the characters are doing the talking and he's just writing down what they say, and I'm eating up every one of his words, because they're impassioned and driven and connected. They're everything Cheyenne shouldn't be, and yet they're spilling out of Cheyenne's mouth as he's eating pizza and walking on the stars.

A PRISONER OF THE PRESENT

Love is so short, forgetting so long.

—PABLO NERUDA, "POEMA VEINTE"

Approaching Julie is like walking into a spotlight. Each time I greet her, I get the impression that she's evaluating me, that I'm the one whose every gesture is under scrutiny. My palms sweat, I fumble around for the right words, I smile nervously, and she smiles back calmly, like she sees right through me. Six years ago, Julie lost her ability to retrieve her memories—all of them. She is so disenfranchised from her past that she generates an unsettling amount of presence, a presence honed by the habit of attention and examination. Her powers of observation are uncommonly fierce, a trait she doesn't acknowledge because she simply doesn't know any different. Julie can't even remember what it's like to have a memory.

When I pass her in the hospital corridor, she pretends not to notice me. It's a type of permission she extends to everyone, the permission to be a stranger. As I draw closer, she keeps me in her line of vision, just barely, to see if I make any telltale signs. She senses how well she knows me by how directly I walk toward her, how long she holds my eyes, the way my shoulders turn when I greet her. If I were to pass her without saying anything, then she might assume she was wrong about me. Maybe I'm just some stranger on the way to the restroom.

As I approach, I make sure to say her name.

"Hi, Julie," I say. This tells her we've met. She doesn't know that we've met dozens of times before and that we've had several hour-long conversations, but she intuits how well we know each other. She has already guessed we're not related, because I'm standing at a sociable distance. She sees how I'm dressed and assumes I'm an employee, maybe a doctor. She's not crazy about doctors, so she throws me a test.

"Hiiii," she says, drawing out the vowel until there's the hint of a lift at the end. She's telling me that she doesn't know my name, in case I don't already know that.

I tell her who I am, and I drop her husband's name so she can get a sense of how much I know about her. I comment on the weather, talk about my family a little, so that she has time to watch my body language and gibe with my prosody. I tell her how my oldest daughter, Cherish, nearly a teen, is already arguing a lot. Julie's arms unfold and she giggles when I tell her that it'll cost her twenty dollars if she wants to hear the juicy stuff. You can have a sense of humor even if you don't have a sense of your own past.

Julie's brain injury is invisible, the scars along her scalp completely concealed by a drape of dark hippie-chick hair. In her forties, she's less concerned about how she looks than about how comfortable she feels, which is why, nine times out of ten, she'll be in denim shorts and a loose T-shirt. Usually she's hugging a spiral notebook close to her chest. Her tinted lenses evoke the seventies, but Julie wouldn't quite understand the reference. The seventies are the same as the sixties and the eighties and the nineties, which are the same as yesterday and a hundred years ago. The past doesn't exist for Julie Meyer, not as it exists for you and me. We can daydream about childhood vacations and graduation, we can describe how we felt at prom, and we can find our way back home after a bike ride. Julie can't do any of those things. Her relationship to the present moment isn't the same as the abiding freeness alluded to in Eastern philosophies; it isn't an embrace of past and future, or an opening to the world's fullness. Hers is an unholy state. Julie is a prisoner of the present.

The past is a place of solace to Rick Meyer, Julie's husband. He longs for the days when he could grab his daughter, Christina, by the waist and swing her giggling toward the ceiling. He misses passing a six-pack of iced beer cans around a picnic table crowded with family and friends. He misses lying down shoulder-to-shoulder with Julie and complaining about how his boss keeps expanding his territory without expanding his paycheck. Rick's life has long been stripped of giggles and picnics and pillow talk.

More than anything, he wanted a child. Rick met Julie in his early twenties and before they even married, it was clear they both planned on having kids, and lots of them. Julie was the youngest of seven children and boasted about her big family gatherings and all the drama their relationships created. Rick's family was much smaller, but he had always been kid crazy. Julie's first pregnancy ended early on account of a blood disease she carried, but the doctors told her that she should be able to conceive again. The next pregnancy ended in a miscarriage, and so did the following. Before the fourth miscarriage happened, both Rick and Julie went through testing at a fertility clinic, but when the fifth pregnancy also failed, Rick began to lose hope.

Toward the end of their seventh year of marriage, Julie's belly began to swell.

"I could tell this time it was going to happen." Rick smiles. "She just kept getting bigger and bigger and I kept getting happier and happier."

We're sitting on some barstools in his dining room, which is decorated in racing paraphernalia and neon beer signs. A corner television is showing a NASCAR race, and I look around and don't see a trace of feminine touch anywhere.

When Christina was finally born, Rick paraded her around the hospital, beaming unabashedly. He was the first to introduce Christina to his best friend, Julie's brother, Uncle Denny, who smiled and poked a finger at her nose. Rick told everyone that the little girl in his arms was his miracle baby, and look at her, of course she was worth the wait. She was perfect, healthy, and, from the first day, as strong-willed as her mother.

Christina's birth, however, marked both a welcome and a goodbye for Julie's side of the family. The week before, Julie's sister Sophie died while waiting in a doctor's office, the unexpected result of a medical complication. The family was finalizing funeral arrangements when Julie went into labor. On the first of November, Julie gave birth, christened her baby Christina Sophia, and then later that same afternoon, hoisted herself from her hospital bed and attended her sister's funeral.

Sophie's death initiated a series of heartwrenching blows to both Rick and Julie. Within a year, Julie's father, a drinking man, passed away from cirrhosis. Four months later, Rick's grandfather, his fishing buddy, died silently in his sleep. Through the losses, Christina would climb onto Rick's lap and lay her head against his chest. She was too young to understand anything about loss or death, until the following year, when Rick kneeled before his daughter and clasped her hands. It broke Rick's heart to tell his baby girl that her favorite pal, Uncle Denny, had gone on to heaven. Christina cried at the news and clutched her daddy tight.

After Denny's death, Rick spent almost a week putting himself to sleep with beer can after beer can, but he reminded himself that his family needed him to be a good father and husband more than ever, that Christina and Julie were missing Denny also. But when Rick's father died of Huntington's disease the following year, he wondered how much more a man could endure. By 1999, he and Julie had lost five family members in five years, and with each passing, they took consolation in each other's company. Neither of them could have anticipated that the following year, 2000, would bring them six deaths in six years. By Rick's count, however, the losses number seven.

Much of what we consider essential depends on memory. We carry a host of suppositions about what it means to be human, but without access to our memories, our notions surrounding identity begin to crumble. Among our more fundamental abilities as humans is our capacity to understand one another, a capacity that hinges on shared experience. When the capacity to access and reflect on those experiences vanishes, the emotional constructs of empathy are altered in unexpected ways. Understanding, when confined to the present moment, blooms into an immediacy that can be disarming, intense, and unsettling.

In the same way you can see a faint star best by an indirect glance, you can learn more about Julie's inner life by watching how other people interact with her. In the hospital cafeteria, Julie often sits at a table with another brain injury patient or with a staff member, and there's always a perceivable level of strain in the pairing. If they can, most people avoid sitting next to Julie because they're afraid of her. She isn't violent. She doesn't pose the slightest physical threat, and she isn't prone to outbursts. Julie frightens people because her presence is almost a perfect mirror. If you cry in front of Julie, she will cry. If you shout at someone else, her veins will bulge in anger and she'll get confused. She is emotionally attuned to you before you say your first word. In the past, one of her doctors called her emotions strange and inappropriate, and he was right. Her emotions are strange and inappropriate because they are so honest. Julie may not be able to relate to your past, but in the present you are naked before her. She is pure empathy, unobstructed by ego.

The Freudian model of self doesn't apply to her. Ego relies on memory in order to regulate a balance between drives, and the superego counts on memory to formulate its ambitions for the future. Julie sits outside the Freudian box; words don't exist for the psychological world she inhabits. In Julie's presence, you sense a space that was once filled with an entire personal history. Without even an anchoring reference to the recent past, a conscientious conversationalist might find it difficult to negotiate even a few minutes of dialogue with her. You might ask her how she is doing, and she will say, "Okay," but you will struggle for the next question to ask. Do you like toast? Are you looking forward to swimming this afternoon?

Julie prefers bagels, and she hadn't planned on swimming or not swimming, because she doesn't recall agendas, at least not in conversation. She isn't looking forward to something that requires her to look back, so your question falls flat in her ears. You can tell a story about the time you went swimming and were too afraid to go off the high-dive, but you have to make your story succinct, a sentence or two, or Julie may not recall the beginning, or even the point. To her, a long anecdote sounds like an arbitrary sentence, a non sequitur that requires her to pretend she's following you. Your life is a filmstrip of which Julie sees only individual frames. She will say "oh" to your story, because she might not know if it is sentimental or funny or frightening.

"You think you can get me a light for my cigarette?" Julie asks me as she's outside waiting for a van to take her to her apartment.

"Okay," I tell her, and hand her a lighter. "But you know this isn't a smoke break, right? You know I could get in trouble for this?"

"You won't," she tells me, and smiles. "I'll tell them I can't remember who gave me a light, and I won't be lying."

Rick liked meeting people. He installed windshields on cars and trucks for a living. His territory ran from northeastern Kansas into parts of Missouri, and along his route he met dentists and welders and bakers and bankers and they all appreciated him. The work itself had a calming effect on him. Removing a cracked windshield requires focus and care, but once a new windshield sits in its frame, a car sparkles as if it just came off the line. New windshields make people feel safe, and Rick enjoyed the look on his customers' faces when they whistled at the work he had done.

On the first of February, Rick arrived at work a few minutes early so he could get a jump on the busy schedule. He was in the garage loading up his truck with a day's full of glasswork when he got a phone call from one of Julie's friends. The wreck was all over the news, she explained. They were flying Julie to North Kansas City Hospital while Christina, their five-year-old daughter, was on her way to Children's Mercy Hospital. Rick jumped in his truck and sped the twenty miles north to Children's. He tried to find out about Julie on the way but wasn't able to reach a single family member.

The one person Rick didn't expect to meet that day was a chaplain. When Rick arrived at Children's Mercy Hospital and demanded to see Christina, the ER nurse told him that he either could sit quietly and wait or be escorted out by security officers. He paced in the far corner of the waiting room and wrung his baseball cap in his hands. When he saw the chaplain emerge from a corridor, his face went white and his eyes welled with tears before the chaplain had a chance to speak.

"It's bad," the chaplain said. He offered Rick a chair and explained that the air crew revived his daughter on the helicopter, and now she was being taken in for X-rays. The surgeons are doing all they can, but she's in awful shape—she's not even breathing on her own. They didn't have any word on his wife, Julie, who had been flown to a separate hospital in northern Kansas City. The chaplain laid a hand on Rick's massive, heaving shoulders and prayed for his family. Across town, a hospital chaplain was having a similar encounter with Julie's mother and sisters.

The seconds passed like hammer blows and the hours moved like glaciers. When Rick thought he would finally shatter from worry, an ER nurse appeared and offered Rick his first glimpse of Christina. She brought him back into a hallway where, at a distance, he watched a man in scrubs wheel Christina's gurney toward him. Her small chin poked out from a neck brace and dried blood caked her arms and face. The tubes going into her delicate hands seemed larger than her own fingers. The man explained that he needed to take her to another room, and within moments, she disappeared back down the hall. Before Rick could protest, a trauma surgeon tapped him on the shoulder and beckoned him to a nearby light panel. She hardly glanced at Rick as she slapped the X-ray of Christina's head on the panel.

"You can see right there." She pointed to the area where the neck met the cranium. "She has a severed brain stem. It's snapped all the way through. Your daughter isn't going to recover from this. You're going to have to make a decision whether to keep her alive or not."

Earlier that same morning, at 9:06 a.m., the freight train heading southwest toward Aspen Road reached a speed of thirty-seven miles per hour. The windows of Julie's Mercury Tracer were blurred by heavy frost, and Christina sat snug in a child's seat in the rear passenger side of the vehicle. Julie's car came to a complete stop at the corner of Aspen and Eighth. A bird's-eye view of the railroad crossing at Aspen Road tells you everything you need to know. Surrounded by farm pastures, trees and greenery fill the corners of the crossing. The tracks cross Aspen east-to-west at roughly a thirty-five degree angle, obstructing the view for both motor vehicle drivers and train engineers.

When a freight train passes a certain point along the track, a switch activates the railroad crossing system at Aspen Road. On February 1, 2000, however, a sequence of failures occurred. The switch failed to initiate when the train arrived. The lights failed to flash, the safety bar failed to lower, and Julie failed to brake in time. The front of the locomotive exploded into the passenger's side of the Mercury Tracer, spraying glass, metal, and blood in all directions.

There are two collisions when a train crashes into a vehicle. The primary collision occurs when the train first strikes. The impact sends the car bouncing several feet from the front of the locomotive. As the car rests on the track, the train then pounds the vehicle an instant later, scraping and grinding the metal remains along the screeching tracks in a shower of sparks. A Class I railroad freight train extends, on average, over a half mile in length. It pulls fifty-two freight cars and hauls sixty-two thousand pounds. The weight ratio between a Mercury Tracer[1] and a freight train is approximately the same as between an empty aluminum can and a loaded dishwasher.

The impact didn't even register in the locomotive's cab. The freight train pushed Julie's car for several hundred feet before an engineer noticed that there was a car antenna tapping against the train's windshield. The engineer immediately engaged the brakes, and it took nearly a half mile to bring the train to a complete stop. In most traffic accidents, braking duration ends in half a second to two seconds. The freight train that wrecked Julie's car required over nine seconds to come to a complete stop, an eternity for the horror-stricken engineers in the locomotive who were blinded to the catastrophe occurring below their windshield.

During a recent family gathering, I had the opportunity to talk to my wife's cousin Barry, who is a biomechanical engineer in the auto industry. Although I didn't mention it, I had Julie's wreck fresh in my mind. I wanted to know why race-car drivers could walk away from exploding, high-speed wrecks but a five-year-old can get killed at thirty-seven miles per hour.

"There are countless fruit hanging low from the trees," Barry explained. "We've designed many safety features that could save thousands of lives every year, but the bottom line is that the public won't buy it. Besides the obvious issues of size and weight, one of the keys to effective safety is having devices that fit each individual driver in their idiomatic posture. Given the huge variety in body shapes, that can only be done with very complex and expensive mechanisms, so engineers are asked to design to generic one-size-fits-all specs, and of course, one size doesn't fit all. As a result, the phrase you'll hear all the time is that a safety device is too restrictive. The five-point belt restraint is too restrictive. The HANS[2] device is too restrictive. Try telling people they have to wear a helmet every time they drive and see if they'll go along. A safe car isn't a comfortable car, and automakers can't sell that level of safety." Most of Barry's time gets spent designing seats that are going to heat people's asses, not save them.

The common driver has but two safety restraints: air bags and seat belts. The ubiquitous air bag still suffers from controversies over its low cost-to-efficacy ratio.[3] Seat belts are required in forty-nine of the fifty states (New Hampshire still resists). Throughout the world, the most popular type of seat belt is called a three-point restraint, in which a single strip of webbing is draped across the shoulder and lap. Currently, only children and race-car drivers are privy to five-point harnesses, in which a seat belt forms an X over the chest and another strap rises from between the legs to strap at either the waist or the chest. The five-point harness is regarded as a far safer seat belt, but it's also more restrictive, with burgeoning body sizes becoming an issue for many would-be beneficiaries.

To date, only one Volvo concept car suggested a significant improvement in seat-belt restraint systems, but like so many other fruit on the tree, it has failed to gain market recognition. Racing-safety engineers anticipate that most automobile safety features in the foreseeable future will center on improvements in braking systems and tire technology. For the meantime, we must resign ourselves to flying down highways at breakneck speeds with only a strap of nylon around our chests, trusting some deployed pillows to cushion our lives. Our heirs will regard us as the comfort-seeking barbarians of the technological age, to be certain.

When I visit Rick at his home just north of Kansas City, he shakes my hand at the door, swallowing up my fingers in his large, sandpapery grip. He leads me to a frame on the hallway wall and points out the people in the photos. There's a picture of him and Denny holding up a day's worth of largemouth bass, a shot of Julie with all her sisters, and another picture where he and Julie look as if they're far too young to be on a honeymoon. Although he sounds like the muffler's growl on a twin-belt Harley, Rick's voice rises a pitch when he talks about Christina, his angel. It's been six years since she passed away, but her memory is clear and unyielding in Rick's mind.

"That's her, on the playground with her cousins," he smiles awkwardly. Christina is wearing a white T-shirt and jeans. She has her dad's dishwater blonde hair and a grin that takes up half her face. While looking at the picture,

Rick tells me that she saved eight lives through organ donation, but it isn't a consolation to Rick, it's just the right thing to do. You give life when you can.

Rick doesn't pass a day without thinking of Christina, and from talking to him, you can tell he's jealous of his memories. He spends his days wandering through his spare three-bedroom home in northern Kansas City. When the television goes off, he lets his mind roam and dream about the times he would get Christina ready for school in the mornings, and he grins when he thinks of how she used to "tear the house apart" while Julie slept. He bathes and swims in his memory, loses whole afternoons to daydreaming.

Julie doesn't remember the train. When she wakes up two weeks after the accident, she wants to die. She can't speak, but to Rick, it's all over her face. She's scared and mad and confused all at the same time. She doesn't have that "last thing I remember" moment because there is no longer an "I remember." All Julie knows at this point is pain. The tube down her throat is gagging her, the catheter running between her legs feels like barbwire, and nothing looks remotely familiar. She cries and cries. It's the only thing that makes sense.

The train wreck gave her a cardiac contusion, a bruise on her heart. Julie endured multiple bone fractures, as well as a seven-inch scalp laceration along her temporal-occipital region, but her skull remained intact. The physical forces at work in the collision twisted and wrenched the synaptic connections so violently that they disrupted untold numbers of neural pathways, resulting in a diffuse axonal injury: diffuse because it is spread arbitrarily throughout the brain, axonal because it causes trauma at the neuron's most vulnerable spot, the axon. Through a microscope, portions of her brain look like a forest of twisted and torn branches. "Shearing" is the verb most commonly used to describe a diffuse axonal injury. In the accident, Julie's ability to remember her life was literally shorn apart.

At the site of the wreck, she was extricated by firemen, intubated by an emergency medical crew, and flown to North Kansas City Hospital by a helicopter pilot. She remained unresponsive while family member after family member filed into the emergency room inquiring about her. Although she had significant bleeding and internal damage, the ER surgeon stabilized Julie before nightfall. After enduring so much head trauma, Julie was in danger of sustaining another brain injury from swelling complications. The surgeon burred a hole just past her hairline on her forehead and set an intracranial pressure (ICP) monitor to alert the nurses if the swelling became uncontrollable. Before transferring her into ICU, they also placed a PEG tube so she could be fed directly through her abdomen. Julie also required a ventilator's help for breathing.

"There were a lot of times I could tell she didn't recognize me in the hospital," Rick says. "I learned pretty quick that she wasn't going to be the same."

While Julie shuffled between rehabs, Rick began the grueling legal battle against the railroad company. His daily life consisted of depositions and affidavits, and of retelling his story countless times. In the midst of the upheaval, Rick tried several times to return to work, but between the tears and the lack of concentration, he found his job impossible. In the three months following the accident Rick lost forty pounds, and he claims he hasn't had a decent night's sleep since. Rick's only solace was his regular visits to Julie's bedside.

It took Julie six months to reclaim her speech and walk again—gains that nobody quite expected, given the severity of her accident. After a time, Rick could hold conversations with Julie, but his biggest challenge was getting Julie to acknowledge him. Julie's family had to introduce her to Rick every time he visited, and each time she seemed incredulous that the man before her was her husband.

"Sure, Julie learned how to walk, then bathe herself, then do laundry," Rick says. "But by the end of rehab, I got a good idea of where she was at—basically the level of a twelve-year-old, or less. I couldn't talk the same way to her anymore."

For a brief period of time following her accident, Julie lived at home, under Rick's watch. As Rick tells it, the experience was a trial of his endurance. Her impairments put her in a perpetually frustrated state. She yelled at him if he tried singing along with his favorite songs, she threw jealous tantrums when his friends came over, and she slammed doors if she didn't like what was on television. One day, while Rick was gone, Julie decided she would leave the house for a walk around the neighborhood. By the time she had walked two blocks away, Julie couldn't recall how to get back to her house. In a panic, she jogged in another direction and eventually flagged down a young boy on a bicycle and asked him if he knew where she lived. The boy pointed to the house across the street, and Julie ran home to her porch. She called Rick in tears and promised she would never leave the house by herself again.

Rick lived daily with the grief of his daughter's death and his wife's lost life. Julie was incapable of mourning a daughter she didn't know, nor was she able to understand the sense of loss Rick felt in her presence. She couldn't just pick up an identity as a spouse, not when she didn't know what the role entailed. For her, living with Rick was like starting a relationship with a stranger. The same was true for Rick. He could handle Julie's agitation and stubbornness, but he realized she was beyond his help when she started calling him with requests to kill her.

"If you can find some way to shoot me without anyone finding out," she told him in between sobs, "please, please do it. I don't want to live like this."

Julie's suicidal thoughts weren't built on a sense of loss from her previous life. They were formed through a lament of her current predicament. Although she had no frame of reference, she could not stand the sense of imprisonment she felt at all times. She knew that grown women were supposed to be free to drive cars and go to jobs and shop for groceries, and she couldn't do any of those things on her own. Her world perpetually vanished before her; life felt like a labyrinth where every turn was unfamiliar, unwelcome, and dismally restrictive.

At the hospital, Julie's life support wasn't ever questioned. Before the accident, Julie had made it clear to Rick that she never wanted to survive a life-threatening accident, but she never filed an advance directive expressing those wishes. Julie lived, a fact that she regrets to this day. Suicide isn't an option for her—people who kill themselves don't go to heaven, she says. In a fit of desperation, Rick bought a separate home for Julie and had it staffed with part-time nurses who would arrive early in the mornings and stay until Julie prepared for bed.

The arrangement only compounded the problem. Typical of many brain injury patients, the change of environment caused Julie tremendous stress. Nothing looked familiar from day to day, strangers showed up at her door every morning and ate her food and watched her television and wouldn't leave. Julie argued that she was a grown woman and took the help as an insult. She openly berated her helpers and complained—legitimately—about her lost privacy. In a matter of weeks, Julie's agitation and verbal attacks became daily threats to her caretakers.

During my initial evaluation of Julie, I asked her a series of memory questions that tested various aspects of her functioning. She could tell me the day, but not the month or the year. She knew the president's name, but started to mutter a few consonants as she fished for the previous president's name. She looked at me as though she expected me to answer, and I almost fell for it. As I asked more questions, I watched her face and noticed her eyes were busy scanning the room with each question. She was desperately mining her environment for clues: she had squinted at a distant calendar when I asked about the date. She looked at the flag when I asked about the president, and she watched my lips for any clue I might give. Julie was playing me. Further testing revealed that Julie suffered worse memory impairment than anyone I had ever met.

Julie's brain no longer processes memory as it once did. Her episodic memories—the vivid recollections that most of us associate with memory—are inaccessible, but she has retained a compromised semantic memory; she can recall the names of a handful of people and a few key dates. Her procedural memory, however, functions just fine. She can cook a meal, clean her room, and make a phone call precisely because her procedural memory is intact. But ask her what she ate this morning, or what color her bedroom walls are painted, and she'll shrug her shoulders. Julie is blind to everything that happened more than a few minutes ago. Without episodic memory to give her a sense of continuity, Julie's actions are divorced from one another. When she attempts to read, the words vanish from the front of the paragraph; when she watches a movie, every scene is the opening scene.

Without a rigid, externalized framework to recuperate in, Julie wouldn't be able to pick up any compensatory strategies—the world before her eyes would remain a fog. It was no wonder she lived in a constant state of frustration. In a brain injury rehab, however, she might learn to develop cognitive supports that could act as a pseudomemory, allowing her to gain a level of clarity and diminish her frustration. Julie would require a customized program where she could retain a certain amount of independence while at the same time undergo intense observation. She wouldn't be able to find a sufficient level of personal attention in a large rehab, and she stood a chance of taxing the personnel requirements at smaller rehabs. Julie wasn't an easy fit in any rehab, and had a high potential for failure wherever she went. After meeting with Julie, I called my boss and told him that we had to accept her case.

Julie doesn't remember moments, but she remembers love. Julie knows that she had a daughter named Christina who died in the wreck, but she doesn't remember a single encounter with her. She knows she has a husband named Rick, but she can't recount a single thing they've done together. Each time I have asked her to describe them to me, her answers change a little. Julie wants to believe she knows about Christina's favorite games and stories; admitting otherwise would strike her as unmotherly. Rick's looks and his history don't matter too much to Julie; it's enough that he's alive and reachable, even if they don't live together anymore.

Although it isn't quite home, Julie tries her best to spruce up her room. At least fifty elephants sit atop the dresser and nightstands in Julie's room, including stuffed elephants, ceramic elephants, elephant drawings, elephant menageries, and elephant jewelry. Among the elephants are pictures of Uncle Denny and Sophie and Rick and Christina. Word-search puzzles, pens, and notepads lie in neat stacks and rows. Julie shows me a picture of herself and Rick standing on a beach. I remember Rick telling me about the trip, so I ask her if she knows where the picture was taken.

"Mexico?" she asks.

"No," I tell her. "Hawaii. Rick took you there last year. You had a tough time being away from home, he said. You yelled at him in the airplane, coming and going."

"Oh, yeah." She smiles. "That sounds like me."

Julie offers me a seat on the bench opposite her bed. I look around and take a quick inventory. Her linens are folded and stacked neatly on the shelves in her closet, and her clothes all hang in the same direction. The room is more than tidy; it has that controlled, obsessive-compulsive quality that I've seen in so many homes of brain injury survivors. Julie, Rick once told me, likes things a certain way, and I heard the clinical translation right away. Obsessive-compulsive disorder is such a common trait among survivors that it's often neglected in a physician's diagnosis. In Julie's case, however, her meticulous ways may actually support a sense of comfort, a visual reminder of order in the midst of uncertainty.

Although I've formed a good idea of what Julie's memory is like in a clinical sense, Julie constantly reminds me of how much we don't know about her memory. What are the mechanisms of forgetting? Do memories of Christina exist somewhere in Julie's head, along with eighteen years of marriage to Rick? What portions of Julie's amnesia are psychological and what portions physiological? Will any of Julie's memories ever return?

I ask Julie, in all earnestness, what goes on in her mind when she thinks about the past.

"It's kind of like being blind, I think," she says. "I can tell you the name of my high school, but I couldn't tell you what it looks like."

"Do you get images in your head at all?" I ask her. "You know, like dreams?"

"I don't know if I have dreams," Julie says, "because I don't remember them."

Today, Julie lives in Brookhaven's Transitional Living Center, a group home for brain injury survivors who are preparing to reenter their former communities. Each day, she engages in intensive therapies that, over time, have helped her reconcile her impairments with her wishes. She developed a sophisticated note-taking system that acts as a memory aid; in a stroke of creativity, she integrated it with her smoking schedule. Each time she reaches for a cigarette, the most important reminders pop up from her pack. Her journal is now an inseparable part of her. It's filled with arbitrary, brief observations: the blue door is locked, I ate chocolate after lunch, the moon is behind clouds. Julie refers to it constantly. She also shops at Wal-Mart, she attends music concerts, and she pals around with several of the other residents. All these seemingly casual excursions happen within a strict schedule, and are often planned for days in advance.

Julie was raised in Catholicism and maintains its core beliefs. Every night, she kneels at her bedside and prays for the people she loves by name. She doesn't pray for anything specific, just that God will keep them safe and will show them the right directions to take in their lives. She prays for Rick and for her caretakers and she asks God to bless the souls of Denny and Sophie and Christina. She believes someday she'll be reunited with her loved ones.

Though she may not be expecting it, there is also memory loss in heaven. In one of the more disquieting passages in the book of Revelation, believers are told that God "will wipe away from them every tear from their eyes" and there will be no more "mourning, nor crying." Such an act can't be accomplished without some degree of amnesia. The prospect of further memory loss in the afterlife doesn't sit well with Julie, but she remains steadfast in her faith anyway.

Julie tells me that she's waiting for the day she can give God "a good talkin' to." She wants to know why He allowed her to live through a wreck that she didn't want to survive. She wants to know why He would take her daughter away, along with all memory of her. She wants to know why she has to live the rest of her life with people always taking care of her. Although she can't remember her sins, she can't imagine that she ever did anything to deserve losing a child, a family, and a marriage.

We place a peculiar spiritual value on memory. Notwithstanding Revelation, the allure of heaven itself depends on memory. Most people act appalled when confronted with the prospect of having no memory of this life in the afterlife, and yet none of us seem to have a problem with the fact that we don't remember our first few years following birth. Even so, a future potential for memory loss looms before all of us as we age. We will forget the weird girl in the cubicle next to us, and we will forget our second-quarter grade in chemistry. In some future dementia or god-forbidden accident, we may, like Julie, forget our first kiss or our best childhood friend, or we may forget the times we begged for death or let down our loved ones. Forgetting is hell, forgetting is heaven.

Still, I'd like to be there when Julie gives God a mouthful. I'd like to hear His excuse.

AN INSULT TO THE BRAIN

At nine years old, I slipped off a skateboard and planted my forehead into my aunt Marie's concrete driveway. I sported a goose egg over my left eyebrow for days and milked my mother for candy bars and ice cream. In eleventh grade, my high school buddy dared me to hold on to the neighbor's hotwire fence; I remember nothing except winning the dare. In college, my brain endured a series of minor assaults, most of them chemical. In the past decade, I have unwisely stepped into a boxing ring, I have narrowly escaped a drowning, and not long ago, I sustained a fever high enough to cause a delirium—I imagined that a small camera crew had snuck into my bathroom to film me vomiting. My brain has been bashed, poisoned, electrocuted, suffocated, jolted, fried, and assaulted. Since I've been a brain injury case manager, I've been more careful, but that hasn't reversed any of the previous insults. There has been a cost, I am certain.

My brain can no longer do the cognitive acrobatics that it performed in high school. I can't recall lines from many of my favorite movies anymore, and I can't spontaneously impersonate actors and mimic accents like I once could. Background music gets to me. I have a slow, almost glacial response time when my wife asks me the most mundane of questions—an unfortunate attribute that perpetually irks her. My eldest daughter ribs me over my lousy math skills, and everyone I know laments my inability to keep more than one item at a time suspended in my short-term memory. Invariably, I will forget the butter, claim you never asked me to pick up your mail or feed your fish, and fail to show up to your daughter's bat mitzvah (though the last may be a selective memory problem). These minor impairments are just a hint at the myriad malfunctions I've noticed in the past few years.

So far, I haven't been knocked unconscious by blunt trauma, despite the hopes of an infuriated few. After thirty-five, I will have narrowly escaped the window in which most males get brain injured, but thirty more years from now I expect to enter a new bracket of high-risk TBIs: senior citizens. Seniors are one of the most underserviced demographics, and not a single brain injury rehab in America specializes in geriatric injuries. Each period of life carries its own various risks for brain damage, but the most honorable response to the threat does not involve the avoidance of the activities you love and enjoy; it requires the diligent and consistent application of prevention methods. Harness yourself in vehicles, on rooftops, and among tree limbs; check safety ratings before you buy a car. Show me some mercy and wear a helmet. After a certain age, call a younger person to change the lightbulb. If you and I make it through this life with only a series of minor knocks and jolts, it only obligates us all the more to champion the rights of those who have taken worse blows.

Prevention has its limitations, though. If you were to adhere to every safety standard, you couldn't leave your bedroom. Brain injuries happen to the risky, but they can also happen in the most subtle and unexpected circumstances. A stroke can take out a healthy mother on her way to daycare. Aneurysms happen in children. And I know, from experience, about the kind of brain injuries that can occur at birth. But there are other brain injuries that happen so slowly that the brain is already compromised before the injury is detected. Parasites brought into the body through undercooked meats can burrow into the brain, creating the kinds of patterns you'd expect to see in an ant farm, not a human head.

Abnormal growths are a slow, menacing threat, but they're not as rare as you would hope. There's a good chance you already have a brain tumor. It's either in your head or in the head of someone you love, so it might as well be you. Every fifth person has a tumor somewhere in his or her skull, quietly embedded in a gland or elsewhere, too small to see and too scary to want to see. For most of us, this tumor will remain still and undetected, and we will pass our lives pleasantly unaware of its presence. This year, however, nearly two hundred thousand Americans will discover that the tumor multiplied, setting off a horrific chain of cell growth. One in five of those diagnosed will receive the worst news: primary malignant brain tumor. Forty thousand of us will learn that we have only a 32 percent chance of surviving the next five years.

A brain tumor's cruelest characteristic is its penchant for children. Most of the people diagnosed with a brain tumor will be under twelve years old. The injustice comes with a bittersweet mercy: at no other time is a person's brain more capable of recuperating from such a condition. As many as 69 percent of children with a brain tumor will survive. Their brains will be monitored, tested, prodded, radiated, or even resectioned. After the chemotherapy and the surgeries and the lab work, the child will be left with deficits, physical or cognitive or both. It is at this point that the young human brain rallies its most mysterious resources and embarks on a hopeful journey toward restoration.

Before hope, however, comes fear.

Ounce for ounce, a walnut enjoys more protection than the human brain. A scant quarter inch of bone separates the inner and outer world, with males usually possessing the denser skull—a grace, considering the gross propensity of males to injure their brains over females. At birth, the cranium is nearly full-size, though rubbery and thin. During the first few months, an infant's skull is so delicate that you can count a baby's heart rate by watching the pulsing fontanel point on her forehead. The cranium's eight plates are initially held together by connective tissue, which later form into hardened articulations, or joints, called sutures. Simultaneously protective and restricting, the cranium is the brain's best defense and its greatest liability in times of trauma.

Lining the inside of the skull is a translucent, latexlike layer called the dura mater that helps prevent the brain from rattling about like a rock in a box. Beneath the dura mater is the arachnoid layer, a wispy, weblike layer that resembles damp cotton candy. The pia mater, under the arachnoid layer, hugs the brain tightly. Together, the three layers form the meninges, which keep the brain floating comfortably in cerebrospinal fluid. Inflamed, the dura mater can choke the brain; ripped or torn, it can leak or bleed a brain dry. Once the dura mater has been removed, however, a coroner can extract the brain quicker than most people can shell a crab leg.

Out of its casing, the brain resembles a miniature boxing glove with two thumbs, one on each side. The thumbs are the temporal lobes; the fingers, knuckles, and back of the hand are the prefrontal cortex and parietal lobe; the wrist is the occipital lobe; and the palm is the midbrain. Its texture evokes funnel cake, but its odor—musky and dank from saturating in cerebrospinal fluid—calls to mind less pleasant images of locker rooms and urinals. Living human brain tissue feels like spongy, firm gelatin—more like cheesecake than crème brûlée, according to one surgeon. Blood flow infuses cerebral tissue with a pinkish cast, but upon death the brain glistens gray.

There are patterns and tendencies and types of brain injuries. Addicts, especially those behind wheels, frequent my caseload. Most of the e-mails and phone calls I get involve motor-vehicle accident cases, and every other referral is a fall, a suffocation, or an assault. Peppered throughout the referrals are strokes, neurological diseases, poisonings, the whichever-ectomies, and the curveball. Once a week, usually while reading someone's file, I come across a new method of injury, one that forces me to set down the stack of papers and take a slow breath.

No two brain injuries are alike, and yet the damage is necessarily categorized and typecast into a treatable framework. Every brain injury involves trauma, and every brain injury is by some means acquired, but the terms are disparate in medicine. Traumatic brain injury (TBI) involves an external force that disrupts the brain; acquired brain injury (ABI) happens when the brain's integrity is compromised at the cellular level, which typically involves oxygen deprivation or the introduction of a toxic substance or infection. Shaken baby syndrome, for example, is a TBI, while meningitis encephalitis is technically an ABI. In many cases, however, TBI and ABI happen concurrently. Trauma occurs to the head, and the brain starves for oxygen as the result of either blood loss or swelling. Outside of hospital walls, however, an injured brain is always traumatic, always a TBI (unless you live in Canada, where it is always an ABI).

When a brain injury is the result of trauma, it falls into two slapstickish subtypes: closed head injuries and open head injuries. A closed head injury is just as complicated and vicious as an open head injury, sometimes more so. During a closed head injury, the brain may slam against one portion of the skull, then bounce against the opposite side of the wall, causing a coup-contrecoup injury, where two injuries occur from one blow. When head trauma is evident but undetectable by medical equipment, the diagnosis is usually a concussion, but when the bruising of the brain is visible, it's a contusion. If the head is whipped around, a tearing effect occurs throughout the brain, resulting in a diffuse axonal injury. Axons are the hairlike extensions of nerve cells that transmit messages, so in a diffuse axonal injury, the messages either get mixed up or don't come through at all.

One of the particularly gruesome dangers of a closed head injury is the tendency of an injured brain to swell. The head is already a crowded place, so if there is no room within the cranium to expand, the brain takes the path of least resistance and starts pushing against the eye sockets. The optic nerve eventually gets pinched, and eyes appear fixed and dilated. Trauma surgeons often burr holes to monitor cranial pressure, but if the swelling is too extreme, the only other alternative is to create an escape hatch by sawing away a portion of the skull, called a bone flap. Sometimes, surgeons keep the bone flap in the fridge, other times, they simply slit open the survivor's abdomen and tuck it under the skin for safekeeping. Storing a bone flap inside the body minimizes the chance of losing the tissue to infection or freezer burn, and it's much less likely to be misplaced. Military docs, on the other hand, just throw away the skull and order a new plastic one to be installed at a later date.

Open head injuries are a frightening mess, literally. Whether the insult comes from a bullet, a baseball bat, or a high-speed collision, the result is always chaotic and distressing. The scalp is so vascular that blood pours liberally from any laceration. When bone is cracked or penetrated, shards invariably get lodged in the brain. The surgical extraction of foreign bodies requires yogic patience and focus. A range of similarly harrowing, risky surgical procedures gives brain surgery equal footing with rocket science in the common vernacular.

No matter the type of injury, the event is often annotated in medical records as an "insult to the brain," a phrase a scientist, not a writer, must have adopted.

"*Insult* is an exquisitely zany word for the catastrophic event it is meant to describe," writes the science essayist Floyd Skloot, himself a brain injury survivor.

Skloot argues that while the etymology of the word suggests assault, the contemporary usage of "insult" betrays our common attitude toward the brain-injured legions. We perceive that the survivor has been wronged, offended, somehow made less as a result of his injury. Skloot has an unavoidable point. Many of the injured carry a sense of negation with them; they do feel lessened, victimized, demeaned, and insulted by their injury. And yet within the same person abides the capacity to be free of the negation. They are not insulted; they are transformed. They are not disabled, but challenged. In their resilience, they integrate their injury into a life rich with experience and fullness.

Although I don't find myself agreeing with him too often, Satan has an undeniably intuitive grasp of the human condition. After a humiliating fall from grace in *Paradise Lost*, he finds himself unable to fly out of hell—it follows him wherever he goes. Bereft and exasperated, he concludes, "The mind is its own place, and in itself / Can make a Heav'n of Hell, a Hell of Heav'n." The devil's wisdom goes double for brain injury. As mysterious as it sounds, the brain is arbiter of its own reality—it sees the world in its own way, and if injured, may perceive it in a fashion utterly alien and inconceivable to our own manner of thinking. Damaged brains can create hallucinations across every sense, and they can create a cognitive disruption so great that you may perseverate on a single word for hours or days. Injured brains manufacture unimaginable realms, forcing survivors to venture into the territory of nightmares and fairy tales. Let the survivor steal some of Satan's resolve. "What matter where," he asks, "if I be still the same?"

ROB RABE CANNOT CRY

Rob Rabe cannot cry. His forty-five-year-old face is boyishly handsome and expressive enough, but he cannot cry. He can smirk and smile and grimace and furrow his brows, but he cannot cry.

"He can cry," his mother, Kathy, protests. "A mother knows."

"I can't," says Rob, shrugging his shoulders.

Despite the volumes that comprise Rob's medical records, his inability to cry isn't even afforded the dignity of a diagnosis; in fact, it isn't even mentioned. Rob Rabe cannot cry, not like the rest of us. He cannot cry and it is a lost detail to his army of doctors and surgeons and therapists and nurses.

Rob cuts an impressive figure. From the confines of a mechanized wheelchair, he maintains his stature, giving you the sense he is towering alongside you, even though he is seated. He welcomed me into a conference room, offered me a lunch of Chinese takeout, and asked me how my trip to Colorado had been and if I'd had any trouble finding his business. "I did have trouble," I told him, "but your secretary straightened me out."

"She does that to me, too," he joked. "All the time."

That week, I happened to be attending a conference in Denver, and when I found an open slot to visit Rob's center in Greeley, I jumped at the opportunity. I had heard Rob's name mentioned several times in the past, and I knew that he was now running a rehab that featured some sort of brain injury program, but I'd never had the chance to meet him in person. I also had an ulterior motive; I wanted to check in on Rob's mother and visit with her as well.

Kathy Herring, Rob's mother, is widely regarded as the matriarch of brain injury advocacy in Iowa—no small accolade, since Iowa has a reputation as one of the most progressive and survivor-friendly states. More options exist for Iowans with brain injuries than exist for most other Americans, and Iowans have Kathy Herring and a handful of others to thank.

When I first met Kathy several years ago, she was a vivacious, spirited woman who was bustling among a crowd of health professionals during a brain injury seminar sponsored by her rehabilitation center, On With Life, located in Ankeny, Iowa. Although it was the first time we had met, she greeted me with open arms and explained that she knew about me and the hospital I work for, and that I was a welcome guest. She's the kind of woman whose warmth and openness can instantly put a person at ease, but there's also a side to Kathy Herring that she doesn't often show, a side tempered through Old Testament trials and tribulations.

The first half of the year had not been kind to Kathy. During Iowa's annual Brain Injury Association conference, the president, Julie Fidler Dixon, announced that Kathy had been diagnosed with non-small lung cancer and that she wasn't able to attend due to complications resulting from her illness. Later, Julie pulled me aside and explained in hushed tones that Kathy wasn't doing well at all, that the cancer seemed malignant, and that it may have spread to her lymphatic system already.

A few months later, I spoke with Kathy on the phone, and my heart sank at the sound of her voice. Gone was the tenacity, and in its place I could make out only a quiet resolve and acceptance. She was on a strong regime of painkillers and struggled to put sentences together, and she complained that she had never known what it was really like to be sick until she had undergone chemotherapy. Although everyone hoped for the best, she told me that she wasn't sure what the next batch of test results would show. "But either way, I'm going to go see Rob in Colorado soon," she told me. "It's going to be wonderful, I know it."

Rob was probably too popular and handsome and promising. At least that's what an old college pal of mine would say about him. "So-and-so is too good for this world," my friend says. "Something always happens to the good ones." And something did happen to Rob. Some say it was terrible, others that it was part of a divine plan, but even fatalism and fate come up short in explaining brain injuries. There aren't any good explanations. There never are.

In 1980, Ankeny was one of Des Moines's lesser satellites, a tiny suburb hovering just north of downtown. It was small and rural, but it held enough allure to Kathy Herring and her new husband, Fran. She was done with Indianola, and a new start seemed just right for the Herrings. Kathy brought her son, Rob, along with her, even though she knew he probably wouldn't be staying around for much longer. He had already graduated from high school and had his eyes set on college and girls and little else.

In high school, Rob had been a jock. Not a dumb jock or an aloof jock, but an entirely approachable jock, the kind of guy that made friends just as easily as pocketing a pigskin into a running back's hands. He was the quarterback, he was the point guard, he was the star. At six feet even and a lean hundred and eighty pounds, Rob felt he was physically unstoppable and had the record to back it up. No wonder, then, that Rob had no idea what to do in the anticlimactic aftermath of graduation. Stocking soup cans on a shelf could buy him only so much time, so as soon as he settled into his mother's house and made peace with his new stepfather, Rob enrolled in a nearby community college and started picking away at his general ed classes—a productive and practical enough decision, given his uncertainty about the future.

Despite his mid-American mannerisms and his aw-shucks smile, Rob isn't normal, not by brain injury standards. A severe brain injury always results in some degree of cognitive impairment, but Rob is lucid, cogent, and clever, with no discernible thought disruptions. He also has a precise recollection of what occurred the day of his injury, and he is the only brain-injured person I have met who can provide most of the details himself.

He went to work. He punched in, he stocked and swept, and he chatted with another stocker about the doe-eyed cashier, Cheryl, who had recently quit. Rob mentioned that he was going to meet her at a party that night and that perhaps he would give him a call to join them, and his coworker shook his head. The jocks always get the cute girls, his expression said.

Rob never did call his coworker, but he did get an early start on the evening. After he clocked out, Rob picked up a twelve-pack and headed to Indianola, but it was a Thursday and his friends weren't around or just not in the mood. So he drove back to Ankeny and called Cheryl, and she gave him directions to the party.

Rob and his coworkers measured time in typical sixteen-ounce fashion and eventually found themselves dancing and drinking, then continuing the fun at a bar downtown. Cheryl talked about her plans to attend Iowa State and become a nurse, and Rob told her that it must be nice to have a dream, or even a plan. As the hours passed, they talked about how they eventually wanted to leave Iowa and make it to a bigger city where the options weren't so limited, and where they could find a little space from their families. They were both ready to move out from under their parents' care, and the more they talked, the more Rob felt sure it would soon be time to leave.

At midnight, only Rob and Cheryl and their coworker Bill were left, so they finished up their drinks and climbed into Bill's Mustang. As soon as Rob sat down, he found Cheryl on his lap, and she slammed the door shut and threw an arm around Rob. The Mustang sped off into the night, heading north back to Ankeny.

There are three different kinds of tears: reflex, basal, and psychic. Reflex tears are the kind that pour out when we chop onions or get poked in the eye—they are the eye's quickest attempt to soothe itself. Basal tears, which flow continuously, keep the eye lubricated throughout the day. Psychic tears are produced by emotional states, and have been the subject of rigorous scientific study, only to remain in large part a mystery.

The chemical composition of psychic tears varies greatly from basal and reflex tears. An analysis initially conducted by Robert Brunish of UCLA revealed that psychic tears actually contain a greater concentration of various proteins.[1] Later research has also noted that the same tears also contain thirty times the level of manganese normally found in blood. Because high levels of manganese have also been found in the brains of clinically depressed patients, scientists have speculated that the shedding of psychic tears may actually purge the brain of manganese and relieve symptoms of depression.

There are several reasons Rob may not be able to cry, but he doesn't know for sure. It could be that his lacrimal system was damaged in the wreck, but that explanation isn't likely. Years ago, I evaluated an older brain-injured woman who also suffered from Sjögren's syndrome, a degenerative disorder that shuts down the lacrimal system. She required an almost constant lubrication applied to her left eye, and her doctor implied that she would soon lose her eyesight as a result of the condition. Rob didn't use any eyedrops during our lunch together, and his eyes were glistening and healthy.

The neurotransmitters vasoactive intestinal peptide (VIP) and acetylcholine (ACh) have been identified as key triggers in the creation of psychic tears. Perhaps damage sustained to Rob's brain stem altered the production or capacity of these neurotransmitters, thereby impairing his ability to generate psychic tears. In other words, even if Rob knew he was sad or upset, his facial muscles might never get the signal. But Rob's face is animated and lively as he speaks, so most likely the neurotransmitters are not the culprit.

Here, as it always seems to, Occam's razor reigns once again. The cranial nerves flower out from the area where the spine joins the head and weave their tendrils upward, around the jaw, eye sockets, and ears, and the back of the head. The most simple explanation is that Rob's cranial nerves were pinched or severed or compromised in just such a way as to let him retain the ability to make one kind of tear, but not the other. If the damage struck a centimeter to the right, Rob's eyelids could permanently droop; a centimeter to the left, and his mouth could pull to one side. Cranial nerve damage is the most likely cause of Rob's missing tears. As it stands, Rob doesn't know why he can't cry and doesn't much care to know. He just knows that he can't, that's all.

Novembers in Iowa are unkind. The air was already icy, the majestic cornfields brown and tattered, and the streets were barren and slow. It was on a mean November night that Wayne Matthews, drunk and running late, raced his way home in his girlfriend's car. He had been drinking all summer and fall, and now he couldn't even afford to pay her insurance as they had agreed. Mad, depressed, and eager for bed, Matthews tore down the quiet streets.

Up ahead, he saw a Mustang. He decided it was moving too slow and passed it before he realized there was a curb on the road. His instincts and fear kicked in a little too quickly, and he jolted the steering wheel to the right. When he merged back into the lane, he clipped the front end of the Mustang, heard a terrible scratching sound, and then a loud bang. The lights disappeared instantly in the rearview mirror and Matthews kept driving, hoping to God that the car had just made a turn and disappeared.

The Mustang did not turn. Instead, it rammed headlong into a telephone pole, effectively splitting the car in half. Rob was dead, Bill was certain, and when the ambulance showed up, the emergency crew rushed to Rob, tried to get vitals, and, failing, pronounced him dead on the scene. The emergency crew bagged Rob, while another technician attended Cheryl. Bill emerged virtually unscathed. A crowd of firemen and police buzzed around, taking measurements and collecting eyewitness accounts.

With all the sirens finally silenced, the scene of the crash grew eerily still as the flashing lights illuminated small islands of wreckage. In that stillness, the ambulance driver turned his head toward Rob's body bag and listened again. He heard gurgling noises, and within seconds he had the body bag opened and was performing CPR on Rob.

"We gotta get him on a flight," he yelled. An EMT grabbed the radio and called in a helicopter.

The Life Flight team agreed to take Rob, but bet among themselves that he would not survive the four-minute flight to the hospital. But Rob did make it and Cheryl did not. Cheryl was a good one, my friend would say. Her father still calls Rob every year on Christmas, just to say hello.

Kathy was the one who answered the door at four in the morning.

"Are you the mother of Robert Rabe?" the policeman asked.

"I am," said Kathy.

"There's been a bad accident—your son is in the hospital," he said. He explained that the wreck had happened after midnight in Ankeny, and Kathy said surely he must be wrong because her son had gone to Indianola, and he shook his head and said he was certain and that she needed to get to the hospital right away.

Kathy and Fran rushed into the emergency room at Iowa Methodist hospital. Kathy told the nurse that she was the mother of Rob Rabe, and the nurse nodded and led them into a small room.

"This is bad," she whispered to Fran. Two years prior, when her daughter was in an automobile accident, the nurse had brought Kathy directly to her daughter on the operating table. This time she had to stay in a small room. When she looked up from the floor, she saw the hospital chaplain approaching.

The chaplain drew a breath and started.

"His leg is mangled, and he's suffered a terrible blow to the back of his head," he explained. "The doctors say there's brain damage, and that there's also a lot of internal bleeding. We don't know if he's going to make it out of surgery."

The chaplain didn't need to ask if Kathy wanted to pray. She had started the moment she heard the policeman's knock, and she would continue to pray at every opportunity. And she would have prayed now, were it not for the nurse's interruption. The nurse explained that Rob was losing too much blood, that they had already gone through the hospital's small supply and they needed more fast. Did she know anyone who could donate?

Kathy spent the next two hours pinned to the phone, calling all of Rob's friends from Indianola. Football players and cheerleaders and teachers and coaches raced to Des

Moines and filled the Iowa Methodist emergency room, eager to give blood.

"He went through so much blood that night," Kathy said, lifting her hands softly to her face. Kathy Herring can cry, I have seen her.

The neurosurgeon finally appeared shortly after seven that morning.

"I'll tell you something right now," he said, raising his eyebrows. "If I'd run across your son like that in Vietnam, we would've just pushed him aside into a ditch and that would be the end of him."

The Herrings looked at each other wide-eyed.

"You're not in Vietnam," Kathy shot back at him. "You're in Des Moines, Iowa, and we're supposed to be in one of the best hospitals in the state!"

The neurosurgeon rolled his eyes and explained to her that Rob had quite a bit of swelling, so he burred two holes in his forehead and affixed a pressure monitor. Rob's liver had been lacerated and his right leg was in pretty bad shape. He measured four points out of fifteen on the Glasgow Coma Scale; the worst rating is three. The surgeon couldn't offer a guess as to when Rob would emerge.

"I think he's going to be just fine," he said. "Why don't you go ahead up to ICU so you can see your son?" The entire discussion hadn't lasted ten minutes and only offered a minuscule bit of information, a phenomenon duplicated and demonstrated thousands of times by neurosurgeons across America. Most neurosurgeons aren't known for their chattiness or their tolerance of family members.

"The neurosurgeon was a jerk," interjected Rob, "but he was aggressive. I've gone back and studied the medical records and he took a big chance with the pressure monitors. At that time, it was incredibly unusual for a surgeon to make burr holes so soon after the injury. I think that's why I don't have cognitive problems, either. I think the guy saved my life."

The neurosurgeon's confidence was almost fatal, though. On the way up to ICU, Rob stopped breathing, and he was rushed back to surgery. Nobody mentioned Rob's turn to Kathy until she approached the nurse's station and asked where her son was. By that time, he had already been stabilized and was once again returning to the floor.

"You're going to need to prepare yourself for this," the nurse said. "It's going to be quite a shock to see your son. He won't look much like your boy."

The night before, Rob had a bushy mane of light brown hair and the penetrating gaze of a strong-willed teen. His chiseled chin and nose complemented the sharp angle of his wide shoulders and taut hips. When Kathy laid eyes on him, she could hardly recognize him. His head had been completely shaved, and it was so swollen and discolored that it gave him the blunt, battered appearance of a heavyweight boxer after losing a prizefight. His eyes were half-opened.

"We can't figure out why his eyes won't close," said a nurse, "but it's really creepy."

"It isn't creepy to me," Kathy said, choking back her tears and smiling at the same time. "He's always slept with his eyes half-opened, just like my dad. I'm used to it."

Shakespeare's King Lear famously wailed, "I am bound / Upon a wheel of fire, that mine own tears / Do scald like molten lead," and his Caliban, after a blissful dream in *The Tempest*, mourned, "When I waked, I cried to dream again."

Alexander wept when there were no more worlds to conquer, Katharine Hepburn could dispense drops at will, and Michael Jordan bawled like a baby upon the Bulls' first NBA title. Neurochemically, all of these tears are the same, but they vary profoundly in their emotional roots. Science can tell us about our manganese levels, but it has to ask if we're happy or sad or both or neither.

This small and seemingly insignificant matter hints at the broader questions currently setting the neuroscientific field afire. If numerous physiological and neural systems are involved in the act of crying, how is it that an emotion can dance between these systems to arrive manifested as a tear? What are emotions, really? Does the brain hold the key to understanding the nature of emotions?

Some say yes, others no, and the arguments are equally rational and inconclusive. For all our brain mapping technologies and consequent research, we do not know where or how an emotion begins or where it goes, and we have had only minimal success catching the traces of an emotion. An entire industry is founded upon the idea that inhibiting the uptake of serotonin relieves depression, but we do not know if our serotonin levels are the cause of depression, or the effect. Science has to ask you, not your brain, whether you are happy or sad.

The Herrings had to make an appointment in order to receive a second encounter with Rob's neurosurgeon, and this time they barraged him with questions. They wanted to know more about Rob's outcome and what part of Rob's brain had actually been damaged.

The neurosurgeon echoed his initial assessment. He told the Herrings that Rob was going to be okay, that he would recuperate well, and that "the sky's the limit for your boy." It sounded like great news to Kathy, so she probed the surgeon more.

"You mean he's not going to be a quadriplegic?" she asked.

"I didn't say that." The surgeon grimaced. He closed the chart and excused himself.

Fran Herring, a former EMT, picked up on one small bit of useful information. Rob's "brain damage" was primarily localized to his brain stem, and it was enough to spur Fran to research. He spent hours in the hospital library while Kathy remained at Rob's bedside. While he pored through medical journals, she endured a seemingly endless parade of students and teachers from every grade, all coming to wish Rob well.

"The boys were peculiar. They got one look at Rob, and that was all they needed. They were in and out, just like that. The girls would stay and get emotional, and we even had two or three faint on us." She smiles. "Good thing they were in a hospital, I guess."

Fran's findings were thorough, but in 1980 brain injury research was still in its embryonic stage, and there was a terrible poverty of information regarding brain trauma. Fran's research was limited to whatever the hospital library had on the shelf and on microfiche, so he decided he would learn as much as he could about the brain stem. From his ambulance-driving days, he knew that the brain stem acted as the brain's major hub to the rest of the body, but he had only a minimal understanding of its functions.

The brain stem is composed of three parts: the medulla oblongata, the midbrain, and the pons. Together, they regulate our basic survival and arousal responses, such as breathing, digestion, heart rate, and blood pressure. One blow or lesion to any of these areas, and a number of atrocities can happen. Even current research is scant involving survivors with brain stem–only injuries such as Rob, because usually other areas of the brain suffer a significant injury at the same time.

When only the medulla oblongata is traumatized, a person can experience either overstimulated effects, such as perpetual, ceaseless coughing, which eventually collapses the lungs, or an eradication of such automatic responses, such as losing the ability to swallow. If the lacerations on Rob's medulla oblongata were too severe, he could be confined to a feeding tube for the duration of his life.

In the event his midbrain was impaired, Rob might find himself still able to see, but without any control of his eye movement. His eyes could wander and cross arbitrarily, and he may not be able to focus or perceive any depth of field. In a worst-case scenario, Rob could lose his vision altogether. Since Kathy and Fran already knew Rob's eyes were crossed, they suspected some degree of impairment in the midbrain.

The pons's primary responsibility involves controlling the respiratory function. When the pons is damaged, breathing may become dangerously erratic or could cease functioning entirely, condemning an individual to life on a ventilator. The Herrings must have also suspected pons trauma as a possible culprit for Rob's breathing problems following his initial surgery. Later reports did indicate a pontine contusion.

At the time, what they did not know might have been something of a blessing. Most severe brain stem injuries result in one of two conditions: locked-in syndrome or persistent vegetative state (PVS).[2] In locked-in syndrome, the individual remains alert and cognitively intact but has no ability to control body movement or function. Communication is often reduced to a sequence of eye blinks, if it exists at all. Survivors have called locked-in syndrome the waking nightmare, and most of them do not emerge from the condition.

If locked-in syndrome is a waking nightmare, then PVS must be akin to a waking death—in fact, it is also known as cortical death and often mistakenly labeled brain death. In a PVS, a person retains noncognitive functions such as sleeping and breathing, but they are presumed unconscious of sensory input or their surroundings. Individuals may laugh, cry, moan, or yell, but they are allegedly unaware of their actions. While a person in a PVS cannot demonstrably respond to most stimulation, they sometimes may react to pain—a basic response endowed to even the most primitive life-forms. Since scientific instruments cannot measure consciousness, many professionals challenge the assumption that PVS is a legitimate diagnosis. The controversy regarding PVS culminated most recently in the media circus involving Terri Schiavo, whose feeding tube was discontinued at the request of her husband. There are rare cases where an individual has regained awareness following a PVS, but some professionals argue that such persons must have been misdiagnosed.

Kathy and Fran had no idea what to make of Rob's injury, but to Kathy, it didn't matter. She never once entertained the thought of giving up on her son.

"I was locked-in and I knew it," Rob said, explaining the emergence from a coma. "It was awful and scary and frustrating beyond belief."

Shirley Flatt, Rob's night nurse, was the first person to notice Rob's eyes darting around the room. He had been in a minimally conscious state for nearly five days without so much as stirring, but lately, between the hours of 2:00 a.m. and 4:00 a.m., his eyes opened and moved and closed and reopened. Nurse Flatt brought it to the attention of the doctors, and they told her not to pay attention to Rob's eyes, that it was probably just a fluke, a meaningless coincidence. Instead, she chose not to pay attention to the doctors and told Kathy about it anyway.

Kathy made sure she stayed at the hospital the following night and set her alarm for two in the morning. When the alarm sounded, she calmly awoke, sat upright, and stared at her son. And she waited.

Rob opened his eyes and looked around. She called to Rob and he looked at her. It was all the hope she needed.

"From that moment on, I started working with him," Kathy explained. "It took us about two months, but once he could keep his eyes open for a while, we would ask him to blink once for yes and twice for no, and we went on like that. The doctors still told me that he was semicomatose and that his blinking was just random. But I knew he was back. I knew my son was back."

Without the aid of any therapists, Kathy began communicating with her son. She would ask him to spell out words by asking him if the first letter was between *A* and *M*. If he said no, then she knew it began with a letter between *N* and *Z*, so she would then ask if the letter was between *N* and *T*. It could take up to ten minutes to spell out a word.

"It was so frustrating I could hardly stand it," says Rob, shaking his head. "Sometimes I would shut my eyes and keep them closed so my mom would leave me alone. But she always knew how to push my buttons."

"A lot of people think brain injury rehab only happens between eight and five," says Kathy, "but it's a twenty-four-hour job, and not enough medical people understand that. I didn't care if I had to push his buttons at 2:00 a.m. or 2:00 p.m., I was going to do whatever it took."

For Rob, being locked-in forever wasn't an option. He claims he simply never allowed the thought, that it would never have occurred to him. He knew he had to get better and he was determined to keep trying. Eventually, the doctors conceded that Rob had emerged from his coma, at least enough to benefit from some therapy, so they wrote an order for a speech therapist to evaluate him.

Her name was Maddie Schoen, and she arrived with the kind of resolve that Kathy desperately needed. After her evaluation, she said that she had just returned from a conference during which she had learned some new techniques for rehabilitating brain injury patients, and she wanted Kathy's permission to try them out on Rob. She had Kathy's trust immediately.

Up until Maddie's appearance, Rob didn't quite have a handle on his condition. He realized he was in bad shape, but after one particular session with Maddie, it finally hit home. She had been trying to get Rob to smile, and he thought that he was complying until Maddie held up a mirror to his face.

"I was completely shocked," Rob said. "My head was shaved, I was cross-eyed, and my cheeks were so sunken I looked like a skull. I had lost sixty pounds. I just couldn't believe that face in the mirror was me."

Given all the time Kathy had spent pacing the halls of Iowa Methodist's ICU, it was inevitable that she made friends with other families. Some of them lost their spouses or their children within a few days, some of them abandoned their loved ones to PVS, and some of them stood by while a doctor placed her hand on the ventilator and turned off the switch. But some of those families remained, and they bonded.

All in all, Kathy made friends with a total of nine families from the time Rob entered the hospital on through his transfer to an orthopedic floor. They all had loved ones who had suffered brain injuries, spinal cord injuries, or both, but they soon found they all had another common denominator: none of them knew what to do once the hospital stay ended. They had all consulted the few social workers available to them at the hospital, and some even dialed a few numbers with the Iowa Department of Health, but the best Iowa could offer them was a nursing home.

"Nursing homes might be the right place for some people, but there wasn't any way I was going to leave my son there," Kathy says, bolting upright. "Not at his age, no way."

In the midst of their turmoils, each family had to face the grim reality that their options weren't just limited. They didn't exist. Most families couldn't even find a nursing home willing to take their loved one, and even if they did, they couldn't provide a substantive level of care. To a brain injury survivor, a nursing home is no more effective than a holding tank. There wasn't a single brain injury rehabilitation center in all of Iowa at the time.

Kathy Herring decided she was going to do something to change that.

"The hospital care was great, and it was awful, depending on who was there," remembers Rob. "For a few months there, I had a nurse who wouldn't change me at night, and

I just sat there in my own mess until morning came. I had no choice."

While Rob recuperated on the orthopedic floor, he rarely missed out on any of the indignities heaped upon the disabled. In his presence, visitors talked about him as though he were as blank and lifeless as a toaster oven. Some raised their voices as though he were deaf, others talked slowly as if his crossed eyes also meant his intelligence had vanished. His friends did what all high school graduates do: they go away to college, they enter the work force, they start building their own families. They get too preoccupied to visit, and they stop seeing the point.

Some of Rob's friends must have been frightened, without a doubt. The young man who had once led them to multiple victories on basketball courts and football fields now lay speechless and immobile and utterly powerless. On top of his limitations, Rob's mind started playing tricks on him. He began to hallucinate and would see snakes slithering from under his bed and up the walls and pouring out of the closet doors. This just wasn't the same guy, they must have thought. This wasn't the Rob Rabe they knew.

Rob's visitors dwindled away until only one friend remained, a buddy whom he still remains in contact with to this day. With the exception of a few close family members, Rob saw fresh faces only on the holidays, and when the faces vanished, it only amplified his feeling of being stuck. He had exchanged coaches and teammates for doctors and nurses, and his entire education was now reduced to studying the contours of the ceiling and the light patterns the windows left on his bedding. The days stretched into months, and the months brought with them small but significant gains. Rob began moving his mouth and shaping words, but the motoric control governing his speech was frazzled and sent air up his nose instead of through his throat. His voice sounded like a high-pitched whine that would have been incoherent were it not for the accompanying movement of his lips.

As early as the summer of 1981, Maddie Schoen started campaigning for Rob's palate lift. It was a new device, she explained, so new that it might not be available in Iowa. Palate lifts are retainer-like devices that are affixed to the roof of the mouth and have a plastic tube that extends down into the throat. Provided the wearer doesn't have a gag reflex, palate lifts then redirect the air into the throat.

"I don't think it'll help him," the doctor said. He looked at Rob, who had been transported to his office. The doctor then reached a latexed hand into Rob's mouth and yanked hard on his tongue. Rob's muscles went into a violent spasm and he shot out of the chair. The doctor looked at Kathy and said, "Well, I definitely can't help him if he's going to act like that." He then turned down Maddie's request for a palate lift.

When Kathy brought it up again several months later, the doctor informed her that because it was an unusual prosthetic with unproven results, her insurance company would not cover it. Kathy called her insurance company the following day, and they said they would cover it if a doctor ordered it.

Another several months later, Kathy pleaded for the doctor to order the palate lift and he replied that the hospital didn't even offer them. He couldn't very well order something that wasn't available, he said.

That was enough to set Maddie afire. She made a few phone calls around the state and learned that the University of Iowa Hospital fitted people for palate lifts, so she made an appointment for Rob right away and promised to secure the doctor's order later. The Herrings didn't have any specialized transportation at that time. Kathy had to lift, push, and jimmy Rob into her car, assemble and disassemble his wheelchair, and then drive two hours away to Iowa City.

When they arrived there, Kathy hunted all over town for a wheelchair-accessible hotel, but there wasn't a single one. They managed to find a motel near the clinic and struggled over curbs and through narrow doorways. It was a labor of joy, because within forty-eight hours Rob would regain his voice.

"I couldn't believe how deep I sounded," Rob said to me, his eyes widening at the memory. "All of a sudden, people could understand me. It was unbelievable."

By then, it was February 1982, a full year and some months after the accident.

The Herrings, along with eight families from ICU, forged the beginnings of what would become a powerful advocacy collective in Des Moines. In 1982 they formed an official group called the Central Area Support Group (CASG) and aligned themselves with another new organization, the Brain Injury Association of Iowa, a tight group of advocates and survivors in Waterloo. Like all support groups, the CASG underwent growing pains. They bickered and argued and formed allegiances and disbanded and reformed. By the time 1984 rolled around, they finally decided to put aside their differences and get to work.

Together with the Junior League of America, the CASG started making inquiries into the needs and numbers of disabled Iowans, and the numbers were shocking and disheartening. Something had to be done. They tossed around the idea of an assisted living center and even formed a task group that petitioned local hospitals and organizations for help and assistance.

"I've never encountered so many 'noes' from people," says Kathy, laughing. "The doctors said it couldn't be done and the hospitals said no way. Everyone we talked to told us it would never fly. Well, for once, they were right."

The support group had been barking up the wrong tree. They realized that an assisted living center wouldn't satisfy the tremendous number of brain injury survivors in Iowa, so they set their sights on starting up an entirely new facility. In 1987 the remaining families formed a nonprofit organization with the aim of creating a brain injury rehabilitation center, a place where survivors could get on with life.

On With Life opened its doors to its first patient on August 12, 1991; it is located just a short distance from the place of Rob's accident a decade earlier. Rob himself has never had the need to be treated at On With Life because he is one of the inspirations for its existence. He has long since moved on with his life.

Rob's hospital-based recovery entailed a now unheard-of forty months of inpatient care, during which he was shuffled between hospitals in Iowa and Colorado. He eventually decided to stay in Colorado and ended up chipping his way through a bachelor's and then a master's degree. Today Rob works for Greeley Center for Independence, an outpatient nonprofit rehabilitation organization that features one of Colorado's few residential programs for brain injury survivors.

It takes Rob three and a half hours to get from bed to work, and another hour and a half for him to retire each evening. He says there are over forty steps a person must take in order to do laundry, and driving around with him, I believed it. It took us nearly ten minutes to simply go outside and get into his black van. Rob insists that if he didn't spend that much time taking care of his body, he would start to waste away. He refuses assistance from his wife, Jeannie, but can't help it if his mother babies him a little.

Kathy, however, isn't visiting Greeley on pleasure. She actually chose Rob's rehab because he could offer her a level of care that she couldn't find in Iowa, or anywhere else in the world for that matter. It isn't every day that a son can pull his mother up from the devastating effects of intensive chemotherapy, but that's exactly what Rob Rabe has accomplished. Kathy is now as strong and impassioned as I've ever seen her. It was difficult to reconcile the feeble voice I had heard on the phone with the energetic woman before me.

I asked Rob—knowing what he knows now about insurance companies and technology and legislation—what would have happened to him if he had been injured today.

"There's no question about it, I would be dead," he said flatly. "I wouldn't have lasted a year."

He explained to me that in today's healthcare system, most people don't have the chance to heal in a hospital over a course of years. According to recent findings, most TBI survivors now average a stay of thirty days in a rehab—a fraction of the time Rob spent in medically supervised convalescence. Rob's candor left me sobered and disheartened; I sensed he was right. For all the limitless potential of technological advances, we live with, and permit, a highly limited level of healthcare, and nowhere is this disparity more vivid than in the lives of brain-injured individuals.

It is a fact that Rob Rabe cannot cry. He cannot cry no matter how many of his patients die, and he will not be able to cry when his mother will some day pass away.

"He can cry," says Kathy. "A mother knows."

PORTRAIT OF AN INJURY

In person, Asya Schween looks like a black and white photograph. Her skin is a wintery alabaster, cast lighter still against her cropped black hair. When I first see her, she is standing outside of her apartment building in downtown Los Angeles and waving at me as I approach in a rental car. She's dressed in a long denim skirt and a frilly white blouse with black polka dots. At twenty-five, Asya has vaults of girlish energy. She comes bounding across the street when I wave to her.

"Michael," she says as though we've been talking all morning, "you should drive straight and make a left two streets up ahead. There is parking behind the building."

"Hi, Asya," I say, glancing at her, then looking at the street. "It's good to finally meet you."

"You have problems establishing eye contact," she tells me. Disinhibition can be an endearing trait of both Russians and the brain-injured. I like her immediately for saying this.

Asya's voice, to Asya, is a problem. It comes out two octaves higher than you'd expect after seeing her self-portraits, but it has such a sweet cadence to it that it isn't at all what she calls "unattractively infantile, repellent, and sadly thick-accented." She tells me that I'll never be able to understand a word she says, and I smile because her English is great, and she is so neurotic that I feel as if I've found a friend right away. Although I'm in California on business, I'm visiting Asya because I want to see for myself the kind of person who lives through a childhood brain injury, earns two master's and a doctorate in mathematics, and then goes on to flourish as an artist.

Like all other Russians, Asya is not quite Russian. She was born in the republic of Dagestan, was adopted by accomplished parents (her father runs a research lab on semiconductors and her mother is an architectural engineer), and grew up in the capital city of Makhachkala, located on the western shore of the Caspian Sea. Asya is a Lak and Laks comprise only 5 percent of the Dagestani culture. Laks, Asya claims, are known for their rigid commitment to children's welfare, which translates into a strict parenting style insistent on scholastic achievement. By the time Asya was ten, she had already been singled out for numerous academic accomplishments in mathematics and science.

Medals are only tin; Asya is more proud of her apartment. Her corner loft overlooks San Pedro Street, on the neglected eastern side of L.A.'s downtown area. A renovated warehouse, the apartment building sits in the center of the barrio, where migrants, artists, the homeless, and any combination of the three congregate. Once you're in Asya's loft, however, you're a world away from the sandy grays and browns below. The walls and the high ceiling are milk-white with large windows meeting at the far corner. The space is filled with enthusiastic shocks of colors, shapes, and textures: a bright orange couch, a giant spherical chandelier made from paper cups, ceramic experiments without clean lines and heavy glaze, shelves of antique medical books, high translucent drapes, and large mounted prints of Asya's work.

"I love taking pictures and the least thing I worry about is the results," she tells me. "When I see them here on the wall, it is like looking at fossils. Sometimes I notice them and they are funny. They are humorous, not macabre. People look at my photos and say what a sad, sad creature I am. It's probably the person looking at the photo that is sad, not me."

I initially met Asya through her art, not her injury. While staying at a trucker's motel in the roadkill-pocked sorrowlands of northern Missouri, I began browsing online portrait galleries, scouring page after page for a face that might offer some refuge. Earlier that day, I had been chased down the halls of a psych ward by a brain-injured biker who wanted to take my head off with his prosthetic limb. His cheeks and mouth were screwed into a stubbled, angry knot; his was a face I wanted to forget. I looked online for other faces—a habit born from discussions with my boss. He's always half-joking that the world is seen only in projection. Only by looking at others, he once told me, are we able to see ourselves.

Through a serendipitous series of links, I encountered Asya's self-portraits. Most of her photographs at that time portrayed her in some manner of distortion. One image presents her nervous head disproportionately large to her bare, feeble shoulders. In another close-up, her cheeks are distended slightly, giving you the impression her mouth is stuffed with cotton. In that same piece, she's holding a wooden doll and looking at the camera with worried exhaustion. Her expressions kept me captive as I sat gazing at her gallery samples. She looked like I felt: raw, undone, shaken.

Over the following weeks, I found myself returning to her gallery. I eventually e-mailed Asya, asking if she would consider selling a print, although she made no indications on her site that her work was for sale. She wrote back, saying that she was in the midst of preparing to defend her doctoral thesis but that she might be able to ship me a print in a few weeks. Pick anything I want, she offered, except for the ones with the scar. I wrote that I would need some time to think about it, and in a postscript asked her about the scar.

Asya had a severe liver malfunction in August of 2003. After enduring five painful weeks in one hospital, three weeks in another hospital, a seven-hour surgery, and then two more weeks of convalescence, she was forced to postpone her doctoral studies until she regained adequate health. "Unfortunately," she wrote, "I can only offer you a photo timeline of that period of my life." She included a few links to images not in her gallery, and I visited them immediately.

The portraits are tender and quiet and lonely, at once reflective of her other work and also a departure from it. The final shot offers a whimsical relief, with her scar pulled into a smile. I e-mailed her back, thanking her for such a personal glimpse into her life and placing my order for a print. I mentioned something about my own involvement with medical traumas, only I dealt with brains and not livers.

"Had I known you were involved with brain injuries," Asya wrote, "I would have told you about mine."

As a twelve-year-old, Asya passed her afternoons cruising the streets near her apartment building on her new bicycle. She challenged her friends to races, sometimes she pedaled lazy circles, and other times she propped her feet up or tried riding without her hands. One day after school she raced alone down an empty street, expecting a free wide turn through the corner up ahead. At the same instant, a dump truck came barreling down the road, giving Asya barely enough time to swerve out of its way. Asya plowed straight into a brick wall, shooting straight over the handlebars. She blacked out when her forehead hit the wall and woke up several minutes later to a pounding headache. The bicycle frame had snapped in two, and the only thing on Asya's mind was to hide it before her parents returned home from work.

She limped back to her apartment carrying the bicycle pieces in her arms and managed to stow away the bicycle on the terrace. Although it was still early in the evening, she crawled into the bunk above her older sister's bed and tried to calm herself, but she was just beginning to experience the onset of serious symptoms. She vomited violently all over her sheets and pajamas and cried openly over the pain searing her head. When her parents finally arrived at the apartment, they saw their daughter writhing atop soiled sheets and mumbling nonsensically.

"They saw the vomit, and they saw I had bruises under both my eyes," Asya told me. "My parents could not understand me because I was not coherent, but they figured out it had something to do with the bike."

When Asya began to experience seizures, her father, whom she describes as a calm Nordic man, rushed to the telephone and called his sister, a pediatrician. He described Asya's condition to her and she advised him that if Asya remained conscious, she would be safe until morning, when they could then admit her to the city hospital. The advice erred on the side of idealism.

As Asya's parents attempted to calm her, she reported seeing an object in her field of vision. Visual disturbances are one of the most common sequelae of brain injuries and can involve double vision, distorted depth perception, strabismus, or outright loss of sight. Asya's vision problems lingered the remainder of the night, she continued to throw up, she developed feverish sweats and chills, and her body yielded to occasional, brief seizures. In her memory of the night, the physical symptoms paled in comparison to her psychological distress.

"I started having terrible fears," Asya said. "I could only think about my bike and how I had broken it. That night I began to have nightmares that I still have to this day."

"The general notion in Russia," Asya tells me, "is that you don't leave your kids in the hospital."

I'm sitting on one end of her orange couch and she's on the other. The whole time we've been talking she's been a gyroscope of motion. Her face is a fountain of expressions and can deliver overwrought eyebrows, impish giggles, and agitated dimples in a matter of seconds. An American doctor would be eager to diagnose her with ADHD, but I remind myself that the Dalai Lama has a face that cascades with emotion, too, and nobody's writing him Ritalin prescriptions.

The doctors at the Russian hospital ran one meager X-ray on Asya and suggested that she stay in the hospital. They predicted swelling of the brain and said she required medical intervention. Her attending physician asked Asya's parents for permission to insert a pressure monitor into her skull, but her parents resisted. They simply didn't trust Dagestani healthcare. The hospital in Makhachkala had a reputation for accepting patients without having the capacity to properly care for them.

"Kids die there," Asya says. "All they can really do is put you in a bed and wait to see what happens. After I spent one day in the hospital, my parents picked me up and took me home against medical advice."

At home, Asya manifested more physical symptoms. Her heart rate and blood pressure fluctuated wildly, and her memory problems became more evident. She developed painful, TBI-related urinary tract infections. The hospital had given her an anticonvulsant, which controlled her seizures but also made her feel even more nauseated. After the third day at home, Asya became highly abusive, swearing and striking out at her sister, her parents, and even the family dog. She began to speak with a terrible stutter. Prior to the injury, she was considered a sweet-natured child, but after the incident she flew into devilish rages without the slightest provocation.

Arbitrary names and facts disappeared from Asya's memory, much to her family's surprise. She could not remember some friends at school, and she could not recall a single children's story—a heartwrenching loss for her, as her father had spent every evening reading her a kaleidoscope of tales from around the world. After she returned to school two weeks later, her grades plummeted. Still mistrusting of local physicians, Asya's father decided a major financial sacrifice was necessary. He made a few phone calls and arranged to take his daughter to Morozov Children's Hospital in Moscow, a two-day train ride away.

Before we presume that Asya's emergency care was the result of bad advice and the subsequent effects of a dilapidated triage system in an impoverished community, we must acknowledge that hers is not an isolated incident. Had Asya's injury occurred in any number of different countries, it is entirely likely that she would have received similar treatment, albeit for different reasons. According to a 1995 study,[1] between 40 and 60 percent of patients treated for head injuries in the United States were not properly treated, and a similar percentage of patients in the United Kingdom received a similarly inappropriate response. Since that report, however, mortality rates for head injuries have been steadily decreasing worldwide. Advancements in civilian neurocritical care are partly responsible for the high survival rates, and these advancements can, in one sense, be traced to a single tragic episode.

In April of 1989, a young finance executive was raped and assaulted while on a jog in Central Park. Trisha Meili became known to the world as the Central Park Jogger, and her case ignited a national dialogue over issues ranging from the racial profiling of her assailants to public security measures. It was the lurid aftereffects of her brain injury, however, that left one man particularly uneasy. The story of Meili's attack garnered the interest of the philanthropist George Soros, who worked with the Aitken Neuroscience Center and the American Academy of Neurosurgeons to identify obstacles in the treatment of head traumas and to promulgate the findings in such a way as would be beneficial to trauma teams throughout the world.[2]

Today the Brain Trauma Foundation offers its findings online and provides various levels of support that are available to hospitals worldwide. The main goal of the foundation is to educate trauma surgeons about the most reliable methods in the treatment of head injuries through a published set of guidelines that outline, in detail, appropriate medical steps meant to decrease mortality rates and improve outcomes for survivors. One of their most important sets of guidelines governs the response of neurosurgeons to children who appear in emergency rooms with a brain injury.

Thanks to the guidelines, emergency room surgeons are far more insistent about the necessity of intracranial pressure monitors for children. Because cranial pressure is now understood as one of the most significant threats following a brain injury, more steps are taken to control the swelling. The guidelines recommend effective medications and the control of body temperature, and, if necessary, intervention by opening the skull. The guidelines also discourage certain methods that had crept into popular usage and suggest a nutritional support regime to maximize the patient's recovery effort.

The case of the Central Park Jogger initiated worldwide reforms that have since saved tens of thousands of lives. In the United States alone, estimates place the number of survivors at approximately twenty thousand per year over preguideline rates.[3] The guidelines are not solely responsible for the increased survivorship, as technology and prevention campaigns have also had their effects, but the numbers are simultaneously heartening and problematic. Since the mideighties, the United States, along with the rest of the world, has seen a dramatic reduction in the number of specialized rehabilitation beds available to the brain injured. Most rehabilitation administrators that I meet are quick to point the finger at health insurance companies, who have collectively reduced the number of inpatient rehab days covered, as the culprits. Most state officials, however, blame lopsided legislation that ignores the special needs of brain injury survivors and instead confuses them with the mentally ill or the developmentally disabled. The frightening reality is that none of us knows the true extent to which the brain-injured population is denied treatment, and we prefer not to know. Knowing, as one health department employee told me, can cost a lot of money.

After a series of tests that were unavailable in Dagestan, Asya's physicians offered her parents a diagnosis that no longer exists in America or England: neurasthenia. The term was first coined in 1869 by the neurologist George Beard, but continues to fall into disuse because of its hazy parameters. Neurasthenia covers a broad array of neurological and psychological symptoms: exhaustion from mental effort, body aches, sleeplessness, an inability to recover by resting, and nightmares, to name a few. While neurasthenia was once a diagnosis primarily handed to war-torn soldiers, the diagnosis may not have been as off-target as we might presume. Since the pioneering work of Alexander Luria, Russian neuropsychology has made a bold departure from diagnoses attributed to localized lesions, and instead places particular emphasis on the dynamics of brain-behavior interaction. In Asya's peculiar case, neurasthenia seems to fit.

Knowing that Asya and her family could not afford to remain in Moscow for treatment, the pediatric neurologists offered her father a do-it-yourself regime of therapies. They recommended memory games for her amnesia, relaxation techniques for her explosive temper, and neuro-optometric exercises for her eyesight. They supplemented the orders with prescriptions for drugs that were not available in Makhachkala—the drugs would require the help of train station employees to pick up and deliver refills for Asya as her family could afford them. Asya's father took the instructions to heart, and when they returned to Dagestan, he devised a systematic method to address Asya's impairments.

Each day, Asya was required to memorize and recite a poem before going to bed. She continued going to school, but this time with the mind-set that she would reclaim all the time she had lost and raise the grades that suffered. Instead of easing her mental exertion, Asya's parents enrolled her in a dizzying amount of extracurricular activity that involved tutors for math and physics. It was during this same period that Asya's mother enrolled her in a photography class, initiating a relationship with the camera that would accompany her into adulthood. The aggressive revolt against her neurological impairments seemed to work. Asya finished the school year with all perfect marks, save one. The following year, she claimed the gold medal for her republic in the Math Olympiads. Academic ability satisfied only a small portion of Asya's cognitive capacity. Asya's parents did not know how to treat the subtle effects of her injury—effects that have become enmeshed in her life and art.

Today the effects of Asya's brain injury are hardly noticeable to most people; the manifestations are more present as mood fluctuations than cognitive disruption. Asya has a volatility that's been known to scare off unsuspecting friends.

During one incident, she was on a bus with her flatmate of ten years, Dmitriy, when an outburst occurred. Asya began screaming and kicking the seat in front of her. It wasn't until she had finally calmed down that she noticed a fistful of Dmitriy's hair in her hand. The attack is so embarrassing to Asya that its memory still haunts her, but it doesn't cause me any more concern than someone complaining of a stubbed toe. Asya's agitated mood simply sounds like mild seizure activity, nothing special. Episodic rages aside, brain injury may have helped further Asya's artistic sensibilities by challenging her on an emotional level instead of a physical one.

Many of Asya's hardships have been met by a wealth of understanding people in her life. Her parents had the wisdom to channel the aggressive energy caused by her injury into achievement. When she finished school in Russia, she applied to the University of Southern California and was accepted into the master's program in applied mathematics. While there, she completed a second master's degree in computational biology toward her doctorate in bioinformatics. She has since finished and defended her dissertation and has a day job working as a researcher in a children's hospital.

"I don't think I will ever be a great scientist," Asya tells me, "but there is nothing I can't be successful at."

Academic study bolstered Asya's mental agility, but one look around her apartment tells you that her art is her salvation. Diplomas do not hang among the art on Asya's walls, only faces. Not a single shot contains a number. There are self-portraits of Asya as a clown, as a demon, as a dadaistic violinist, as the intimate of a broiler chicken. At times ridiculous, terrifying, worshipful, and deliberate, Asya Schween's portraits reveal a life rich and varied. For the hundreds of children's stories she lost as a child, Asya has created a new canon of fantasies.

After a pleasant time chatting with Asya in her loft, we left to meet Dmitriy for a drink in a downtown café. Dmitriy is also a mathematician, and sometimes appears as a chiseled, stoic presence in Asya's photographs. I was surprised by his in-person mildness and humility—his face hardly changed its kind expression, and he spoke barely above a whisper. We talked a little bit about his research job until a barista called out our order.

As Asya picked up our drinks at the counter, I had a chance to ask him what it was like living with an artist.

"She is the most difficult person I have ever known," he said, in a heavy accent.

He cracked a smile and shrugged his shoulders slightly, and I smiled back.

Dmitriy doesn't work among the brain injured, but I know what he means. Asya Schween is disinhibited, moody, impulsive, maddening, spontaneous, peculiar, brilliant, and uncontained. She's perfect.

THE ONLY THING THAT WORKS

Violence goes with the territory; I learned that from the start. My first encounter with a seriously violent brain injury survivor was by accident, and it occurred before I began working as a brain injury case manager, back when I was a psychiatric technician at another hospital. I wasn't prepared to meet Ben, as he was transferred to the ward in an emergency situation. All I knew was that he'd had a history of institutional placements and that he'd been found wandering the streets miles from his house.

When I buzzed open the door to let Ben inside, I found myself facing a mammoth of a man. My neck craned up to look at him. He wore a blue, triple-extra-large hospital shirt that fit him like a halter top, and he stared straight ahead, not even noticing me below him. A strand of milky drool hung from his lip—he had been sedated to the gills. I should have taken that as a sign, but I was too green. I pulled the door open as wide as I could and Ben lurched inside. His mother, a frail woman in her sixties, led him by the elbow and down the hall to the chairs I had arranged for us.

"I know it's late," I explained to them, "so I'll try my best to make this as quick and easy as possible."

Ben didn't turn to look at me at all. He just kept his eyes fixed straight ahead, focused on a point far beyond the hospital walls.

"Ma'am," I said, handing the form to Ben's mother, "if you could just fill out the top box, and initial those three blanks." She perused the form, which gave me a moment to steal another look at Ben's face. His pumpkin-size head had a Cro-Magnon look to it: a protruding jaw, deep inset eyes, large ears, a dented forehead. He was sitting so still, I couldn't even hear him breathing. I glanced down at his arms and instinctively recoiled in my chair.

"What happened to your arms?" I asked Ben. His forearms were covered with dozens of fresh scratches and cuts. They looked as if he'd been attacked with a switchblade.

"I got mad," he said, finally looking down at me. "I threw the thing in the bathroom."

"He tore a urinal clean off the wall," his mom sad flatly, not looking up from the paperwork. "And then he flung it and it shattered everywhere. It made an awful mess."

I looked at Ben, then turned to his mother.

"Is Ben ever violent toward others?" I asked her.

"Oh, honey, you don't know?" she said. "He'll come after you and the only thing that works is to scream for Jesus."

. . .

Several years after I met Ben, I began working for a different type of mental health center. Very few psychiatric facilities in the United States have neurologists on staff, and even fewer have brain injury rehabilitation programs within their doors. Brookhaven Hospital is different. It is housed in a one-story, inconspicuous building that sits a hundred yards back from Garnett Road in east Tulsa. The locals know it as a short-term psychiatric hospital, and that's a good thing. Most Tulsans don't know that Brookhaven has a nationwide reputation for successfully treating the country's most difficult neurobehavioral cases, and most Tulsans wouldn't want to know what that entails. Typical brain injury programs emphasize physical rehabilitation, occupational therapy, and cognitive therapy. In order to make the waiting list at the Neurologic Rehabilitation Institute (NRI) at Brookhaven, a brain injury survivor has usually exhausted all the available resources in their home state, and they've developed a severe, co-occurring psychiatric disorder. We end up accepting the patients that nobody else can handle.

The NRI wing is tucked in the back corner of the hospital, a short walk from my office were it not impeded by a series of magnetically locked doors, the last set of which are made of steel and built to withstand the same amount of pressure a heavyweight punch delivers. They've been knocked open before. At any given time, orderlies dressed in street clothes dart from room to room along the short hallway while a bustle of nurses, therapists, and doctors rush in and out of the nurse's station at the far end of the hall. There are only nineteen beds available on NRI, and they are nineteen of the most sought-after spaces in healthcare. If caretakers succeed in placing a patient in NRI, it's usually a cause for celebration; it isn't uncommon for a family member to cry when I call them to finally schedule an admission date.

Before a patient steps foot in the hospital, their medical records have been scrutinized several times over by core clinicians so that a treatment plan is already in place upon arrival. I conduct preliminary psychosocial inventories and testing prior to admission, which help the rest of the team connect with the human side of the patient. They want to know what the patient enjoys doing, what they were like before the injury, who cares about them, and what they've done in order to get rejected by all the other facilities. I will tell the dietician to grind the chicken first and I will tell the neurologist to avoid the haloperidol and I will tell the techs that she likes her hair done up on Fridays. The program director will invariably ask me about the left jab or the hair-pulling, and whether she should hide the doorstops again. I might tell her to try the Ramones if all else fails.

The first few weeks of a brain injury program are usually a period of intense observation combined with a scrutinizing method of medication management. Different drugs induce responses at different intervals, they each produce unique side effects, and they all have varying interactions. Finding the right medications for a brain-injured patient is a chess game in which the stakes are lethal if there's a wrong move.

Shortly after the adjustment period, each patient attains a certain degree of celebrity on the unit. We have housed confabulators who habitually concoct any response but an honest answer; we have admitted borderline patients who will attempt to swallow anything within reach; and we have treated violence on a level rarely imagined by outsiders. Each TBI survivor is met with equal dignity and acceptance, attitudes they probably haven't encountered in years. The atmosphere of equanimity is in part attributable to the program director, a down-home, spitfire Southern woman with an endless supply of energy and resourcefulness. Although Pamela Washbourne stands barely over five feet tall, she commands reverence and adoration from staff and patients alike, resulting in a tremendous sense of loyalty and family on the unit, a dynamic difficult to duplicate in larger environments. Charisma aside, it is Pam's innovative approach to neurobehavioral treatment that prompts phone calls from incredulous rehab professionals all over the nation. She claims her secret is simple: let the patient teach you what he needs.

Once you arrive on the unit, you're in a completely different world without realizing it. Every externality is a suspected triggering device for unpredictable behavior. The way a particular vent rattles when the air blows may sound like a jet engine to one patient. Another patient might have such overstimulated tactile sensation that a warm shower feels like a spray of needles. It takes the treatment team weeks of intense observation to figure out each patient's particular triggers, and the solution typically involves an alteration of the environment. Depending on the combination of patients, we have dispensed with doorknobs, worn raingear indoors, removed all unanchored objects, and carried kickboxing pads as protection. Creativity yields uncommon outcomes, my boss likes to say.

Brain injury rehabilitation occurs at a snail's pace, and it works by maintaining a militaristic adherence to consistency. The program is exactly the same every day. You shower, eat, and dress at the same time. There are no surprises and absolutely no deviations. If you're going to the grocery store, you know about and prepare for the trip days ahead of time. Whereas once your internal psychological world fluidly adapted to the world around you, now your outer world must acquiesce to the demands of your impairments. Your new environment is meticulously ordered and rigid, a counterforce to the internal chaos of an injured brain. As uncomfortable and frustrating as the program must be for the patients, monotony and rhythm are fertile climes for the human brain.

In a best-case scenario, a patient comes to us in the throes of frustration, agitation, or psychosis, and over the course of several months, perhaps years, their unmanageable behaviors stabilize so that the potential for harm is diminished or eradicated. The patient then steps down from the intensive neurobehavioral unit and into one of our transitional living centers. Their time in a group home setting is a trial run for real-world reintegration. Most of the patients who succeed in transitional living are able to return either to their families or to appropriate housing in their home states. I have seen patients come to us demonstrating dozens of homicidal intents a day, only to be discharged back to their homes and into the workforce following their treatment.

The odds are slim that a caller will be accepted into NRI. Many of the beds are occupied for years at a time, as are many of the nation's brain injury beds. Only the most persistent and lucky caretakers get their patients accepted into NRI, but the majority of callers will be denied admission. That's where my real job starts. A person calls me looking for a bed, and I tell them we're full and that we have a waiting list. Without trying to sound the least bit hopeful, I ask for a name, date of injury, and an address. Although the caller is completely unaware of it, I'm asking myself if I need to visit the patient, and I'm mining my calendar to see when I'll be within a day's drive of their city. After taking a few notes I set my pen down and put a hand to my forehead. I lean a little into the receiver and I ask them to back up a little. I ask about life before the injury, and a small laugh brightens the line, and they tell me how a wedding was planned or a baby was born, and then this. Their voice gets quieter, more serious and hesitant, and I invite them to keep talking.

Prior to Ben's arrival on the ward, an ER doctor had given him an antipsychotic cocktail that could lay out a water buffalo, but I've seen those drugs wear off. Given his history of agitation and his tremendous strength, I felt a distinct sense of urgency to complete Ben's admission, but his mother didn't catch my vibe. She opened her wallet and offered me a photograph.

"Here's a picture of Ben when he was in school," she said. In a sportscoat with his hair neatly parted to one side, Ben looked like a young, intelligent Republican—he was totally unrecognizable as the giant next to me. His mother explained to me that shortly after high school, a crazed coworker had assaulted Ben. The doctors said that if Ben had been kicked in the head once more, he would have died. Following the attack, Ben no longer knew who he was, where he was, or what day it was. He lived and somehow functioned in absolute, perpetual disorientation. From the looks of things, his injured brain probably hosted a tumor that was causing his freakish gigantism, laying insult upon insult. When a guy gets left in an institution long enough, little things like adenomas get overlooked.

Ben leaned over and croaked a whisper to his mom, who asked me if I could show Ben the way to the gentlemen's room. The closest restroom had three urinals in it, so I decided it would be best to take him to the staff restroom, where he would only have the option of one toilet. I unlocked the door, let him in, and walked away.

Ben let himself out of the restroom and joined us a few minutes later, right after I finished cataloging and securing his few belongings. As he sat back down in his chair, he weaved a few times from dizziness. I told his mother that it was probably best to get Ben to bed as soon as possible, and she agreed. We walked him to his room. He went straight to his bed and lowered himself down until the bedsprings creaked under his weight. His mother pulled the massive no-skid socks off his feet and helped pull his legs onto the undersize mattress. I asked Ben if there was anything I could get him.

"I like to dance," he said. "I like to dance."

"I'll bet you're a good dancer," I replied.

"It's in my heart," he said, curling up like a child. He was asleep before his mother had a chance to pull his sheet over him.

After answering a few minutes' worth of caretaking questions, I let Ben's mother out the front door and told her she was welcome to call in the morning to check on him. As calmly as possible, I ran back to the nurse's station and asked the charge nurse to arrange the appropriate safety measures in the event Ben woke up agitated. Within a few minutes extra staff was ordered, an emergency sedative was prepared, and five-minute room checks commenced. Ben was sleeping safe—from others and from himself.

Before leaving for the night, I walked back down the hall to the staff restroom to relieve my own nervousness. When I opened the door and flicked on the light switch, my eyes were immediately drawn to the toilet seat. The back portion of the rim was smeared with dark, almost black blood.

Pink water clouded the bowl, and the floor glistened with dozens of tiny red dots. I thought about Ben's chart and couldn't recall anything about rectal digging, hemorrhoids, or prostate cancer.

I reported the blood to the charge nurse and expressed concern that Ben might have once again been the victim of something terrible on the streets, and we both looked at each other silently. Hers was an expression that I've encountered dozens of times since, a look that admits that there is no explanation for something like this, a look of utter powerlessness. The nurse reached for her latex gloves and told me that she would take care of it, and that I had done my job, and she would now do hers.

THE RESURRECTION OF DOUG BEARDEN

When he goes dead, his body stiffens and his arms don't move when he walks. In his fifties, Doug Bearden has the all-knees-and-elbows build of a teenager, and cheeks fuller than the past year should allow. In death mode, however, his demeanor strips him of all youthful semblance. His face is stuck in a pale, vacuous blank. Although it makes no sense, Doug thinks he is dead and can't be convinced otherwise, even after a large breakfast and a shower. This is death, he thinks, and there isn't much to do. He paces a few times between the bedroom and kitchen, then decides to take a nap. While he sleeps his wife, Cindy, hopes the feeling of death leaves him, at least for a few hours. Sometimes the death delusion will hover for weeks, other times it may last only a few hours. To her, Doug's death trance is a persistent nuisance, a nagging feeling that doesn't respond to reason or logic. She's tired of it, as tired as Doug must be.

Death is a sensation. It can be felt in the same way life can be felt. Life feels like water and movement and fullness, and death feels dry, like timber and monotony. You and I take the feeling of life for granted; we assume it is a birthright, and that we will own that feeling until our last breath. The sensation of life can be dislodged and lost, though, the same as vision or touch. It can be banged away in a minor car wreck or eaten away by a common virus. For Doug Bearden, the sensation of life isn't lasting, and curiously, wondrously, neither is the sensation of death.

In 1977, Doug Bearden felt more alive than ever. He had recently married both the military and a daisy of a wife, Cindy, who joined him at his new station in Panama. His overachieving type-A personality had just won him a position as an air traffic controller for the United States Air Force, the most stressful job in the world at the time (today it ranks fourth). With only primitive radar technology, scratchy radio contact, and rolls and rolls of air maps, Doug planted himself for long hours atop the control booth at airstrips and guided the endless liftoffs and landings of fighter pilots and cargo planes. Lives rested on his confident calculations. Any indication of stress was an indication of weakness, a hint of error where no errors were allowed. Instead of managing his stress, Doug buried it, until a small eruption near his upper lip gave him away.

He had experienced outbreaks before, to be sure, but this cold sore nagged at him until he decided to have a military doctor take a look. The doctor scribbled down what any doctor at the time would have written: "HSV-1,[1] 2cm, Betadine ointment, no shaving 10 days." Doug simply needed to turn in a little early and stop touching the small, crusting pustule, and hey, keep those lips off that cute wife for a night. The sore vanished from his mind and from everyone's concern, but the virus remained in Doug's body, latent and quiet as a time bomb. Most of us carry the HSV-1 (herpes simplex virus type 1) in our bloodstreams without so much as a blister. Millions more are haunted by reappearing cold sores. In Doug Bearden's case, HSV-1 hid forgotten for twenty-three years, and then, inexplicably, it localized in the most vulnerable place imaginable: Doug Bearden's brain.

The living dead have long been with us. Clairvius Narcisse, the world's most famous zombie, "died" in 1962 in Haiti's Albert Schweitzer Hospital. Twenty years later, he resurfaced as a Baptist squatter and achieved immortality through the publication of Wade Davis's zombie bible, *The Serpent and the Rainbow*. Narcisse's story starts where all good mysteries begin, with a poisoning. In 1962, Narcisse was a deadbeat dad several times over, and more than one woman wished him dead. Voodoo culture permeated the island, so it wasn't difficult for one of Narcisse's more enterprising exes to hire a *voudoun*[2] priest, or *bokor*, to execute the task. With a little puffer fish extract, eau de marine toad, hyla tree frog sweat, and some human remains, the *bokor* whipped up a neurotoxic cocktail that could send an elephant crashing to its knees. It froze poor Clairvius Narcisse to the bone. Within a matter of hours he was stiff enough to have a sheet pulled over his face and, soon afterward, buried in a shallow island grave.

There are Haitian zombies and there are Romero zombies. Director George Romero's rotting zombies shuffled their way to the big screen in the late sixties in *Night of the Living Dead*, a farcical zombie thriller where the quasi-dead proselytize locals by eating their brains. Haitian zombies, however, are characterized by process rather than transmogrification; they are coaxed into a deathlike spiritual passage by a *bokor*, drugged into paralysis, buried, and then resurrected—presumably without a soul. Following their resurrection, zombies become social pariahs in their half-life state and eventually continue on as excommunicated loners who wander the tropical nights. The listless, weak-willed Haitian zombies are created by culture, defined by ritual. They think they are dead because everyone else thinks so.

"Even as they cast dirt on my coffin, I was not there," recalled Narcisse. "My flesh was there, but I floated here, moving wherever. I could hear everything that happened." [3]

Disappointingly, Narcisse's words aren't terribly different from what you might hear someone say in a frat house the morning after a hazing. In his account of Narcisse's in-death experience, Davis argued that a poison called tetrodotoxin caused the stupefaction of Narcisse, which, applied topically as part of a *voudoun* ceremony, made him appear dead to the world and himself. Unresponsive but conscious, Narcisse maintained a catatonic state long enough to receive burial. Davis's zombification theory was eventually drilled lifeless in two science journals, but his account of Narcisse's voyage through the *voudoun* netherworld remains a matter of anthropological intrigue.

Twenty years passed by in the small village of l'Estère, the story goes, before Narcisse showed up again. He claimed he had served time in zombie camp upstate, where he was forced to do manual labor for a bad *bokor*. When another zombie murdered the *bokor*, Narcisse fled and eventually returned home to an icy reception that included a few stones lobbed his way. Nobody, it seemed, wanted anything to do with a dead man. Stripped of cultural acceptance, Narcisse resigned himself to bedding down at a Baptist mission, with only occasional, strained visits home to l'Estère. In Haiti, as well as the rest of the world, the living dead stand to lose more than just their feeling of life; they lose their connection to the living.

Doug Bearden wasn't himself. Each year, the entire extended Bearden family rented a cabin on West Point Lake, just an hour's drive from their home in south Atlanta. Doug usually played the role of crowdherder to the Bearden clan and their friends. He roused everyone early for fishing, he organized the Jet Ski races, and he manned the grill in the evenings. He saw his familial service as a gift to all the people he loved, and it was also a much anticipated getaway from his morbidly hectic schedule. This summer, however, was different. Doug slept in, he avoided conversation, and he even turned down a few challenges at volleyball, preferring instead to sit alone in the shade.

By 2003, Doug had already taken an early retirement from the Air Force, but he hadn't left any of the stress behind—instead, he courted it. He now held two master's degrees, one in education and the other in theology. In the mornings he kept a rigorous exercise routine, then went to work as a regional training manager for a local pizza chain. Evenings found him either coaching his daughter April's softball team or rolling strikes at a nearby bowling alley, and on weekends he copastored a church and made sure to arrange a date night with Cindy. To Doug, West Point Lake represented a return to his original heart, to his family.

"I remember we couldn't get him to do anything," April told me. "It was so strange, because he had been talking about it for weeks. He just sat on a lawn chair most of the time." In her twenties, April is a lithe, sparkling mix of Cindy's southern determination and Doug's affability, but when she's not speaking in a sweet drawl, she wears a tight jaw, bites her lower lip, and kicks a restless leg. She wasn't always this way, she claims.

While the rest of the Beardens enjoyed the water sports and late-night movies, Doug preferred to clam up and turn in early, claiming he wasn't feeling well. A short time after the vacation ended, his slight fever and nausea confirmed everyone's guess that he was sick—a wild guess, as he never seemed to get ill. His body temperature rose along with the blistering humidity, and he complained of muscle aches along his back and neck. Doug forced himself to take a few days off work to recuperate.

By the following weekend, Doug's nausea had subsided but he still maintained a high fever. At one point in the evening, he went around to each air vent in the house and sniffed at the airflow. He told Cindy that he suspected a gas leak, but she didn't notice a thing. She encouraged him back to bed but heard him wake again in the middle of the night.

"Do you smell that?" he asked her. "I think there's someone smoking pot on our front lawn."

Doug jumped from the sheets and peered out the windows. Cindy rolled over and told him to please stop acting weird. It wasn't funny anymore, and it was even less funny at two in the morning. She didn't guess that the smell hallucinations were malfunctioning olfactory nerves, one of the brain's first indicators of neurological crisis. Doug went back to bed and this time slept in late. By Sunday evening, the flulike symptoms had already subsided, but Doug's weirdness hadn't. He continued sniffing at the vents even after April had returned home. She didn't smell anything either.

Although Cindy discouraged it, Doug planned to go back to work in the morning. He had an important nine o'clock meeting to attend at one of the nearby franchise locations and planned on getting up a little early to prepare. Doug went to bed around midnight, but before he did, he asked Cindy if she had changed the time on all the clocks in the house. She giggled at him, told him to stop acting so silly and go to bed already. Doug turned in at midnight on August third. With the exception of a few strange comments, it was the last worry-free night the Bearden family had together.

Several years ago, while I was attending a conference on disabilities in Albuquerque, I met a young mother who had lost her sense of life in a car accident and didn't regain it for a full six months.

"I can tell you exactly what being dead feels like," she told me. "It feels like you're watching a movie of your life, like the things you see happening aren't happening to the real you. It's like being trapped in a dream."

The entire *Tibetan Book of the Dead* is based on the premise that death feels like a dream. Commonly misunderstood as a funereal grimoire, the work is a practical, step-by-step manual utilized by Tibetan Buddhists preparing for the experience of death and possible rebirth. Nightly dreams are welcomed as an opportunity to practice death—the goal is to master lucid dreaming. If one can achieve clear awareness in a dream state, as well as other stages of sleep, then the chances of attaining enlightenment become exponentially improved. More than any other kind of yogi, a dream master is supremely prepared to navigate the three death *bardos*.

According to Tibetan teachings, we experience a profound, sensory dissolution in the moments before death. Upon clinical death, we enter the first stage, called *chikhai bardo*, wherein we meet with the pluralistic bright light. Tibetan cosmology proposes that this encounter is one of the most powerful opportunities for attaining enlightenment. If you can wake up, just as you are able to during a lucid dream, then you may be able to recognize this light as your own enlightened nature. You are the light. Most of us will fail to identify with the divine light, however, and perceive that we are separate from it. We become enthralled with its peaceful radiance and imagine that it is calling to us. The light changes colors, and we are eventually lulled into a still, actionless state before we drift into the second *bardo*.

The *chonyid bardo*, or Bardo of Experiencing Reality, is akin to suddenly waking up, but without your body. Naturally, our response will most likely be disorientation and fear, since our arms and legs and breath and mass and motion are all missing. We are left alone with our projections, which are so powerful they seem to fill the entire universe. This is the stage where we reap what we sow. Our inner lives, our deepest thoughts and desires, fill our perceptions and we encounter them full force. Here we may encounter the deities to whom we attribute our salvation and judgment. The experience may be rich with warm feelings or nightmarish beyond earthly understanding, and yet the aim during this *bardo* is still to wake up and realize that these encounters, too, are illusory. In a sense, we are mistaking the images on the screen for the movie projector, and once again, missing our opportunity for enlightenment. The projections take us for such a sweeping ride that we hardly notice ourselves emerging into the third *bardo*.

In the Bardo of Becoming, *sidpa bardo*, it will finally dawn on us that we have actually died, a true bummer of a *bardo* moment. We may attempt to reengage with the living and might even enjoy our new ability to immaterially manifest in whatever location comes to mind. The earthly world casts a seductive appeal upon us, and if we fail to realize our true nature in this *bardo*, we will find ourselves pining for another round of the human experience. Most of us won't have the presence of mind to choose which womb we will call home, but a few of us may be able to tip the scales for a favorable condition.

If Doug Bearden occupies a *bardo*, it would probably be a distorted version of the third *bardo*. He concludes that he has died, and his mind is obsessed with trying to figure out how his death came about. He paces wildly from room to room, ruminating over his death. He will ask his wife the same set of questions repeatedly. Am I dead? How did I die? Who killed me? Cindy will answer the same each time.

"No, Doug, you aren't dead."

"You didn't die."

"Nobody killed you."

The answers don't make sense to Doug. When he feels dead, the condition haunts him. He will try to lie down and rest, but thoughts of his own death torment him. He can interact with the living, but each manner and gesture reflects an impossible psychological divide. In his mind, every exchange lacks understanding; the words of the living are nonsensical. He will pace the house at all hours of the night, moving anxiously, mumbling the same phrase over and over: I'm dead, I'm dead, I'm dead.

. . .

On the fourth of August, Doug left the house at six in the morning in order to get to a nine o'clock meeting eight miles away. When Cindy asked him why so early, Doug simply told her he was going to stop for breakfast before work. He walked out the door, got in his car, and drove away. When Cindy noticed that he had forgotten his satchel, she dialed his mobile number but heard his phone ring inside the bag. She shook her head, unable to figure out how Doug managed to be so forgetful, a trait he never manifested before. Cindy tumbled back into bed for another half hour, then rose and went to work herself.

She had been at the insurance agency for only about an hour when she got a call from Doug's coworker Mitchell Carson. Mitch and Doug had become fast friends in the past year, developing a strong working relationship where each relied heavily on the other. When Doug hadn't shown up for the meeting, Mitch grew concerned and called Cindy.

"I was wondering if you know where Doug's at," Mitch started. "He was supposed to be here a half hour ago, and it's just not like him not to call or anything."

Cindy was surprised and immediately worried. She thought the worst thing that could happen to Doug was a car wreck, so she began calling police stations and hospitals. Mitch told her that he would send out a fleet of delivery drivers to patrol the nearby streets, and he would be in touch.

A half hour later, a relieved Mitch called Cindy again to say that Doug was on his way in to work. He had stopped in at the wrong location for some reason and was confused when nobody arrived for the meeting. The manager at the other store said that he should be arriving any minute. Cindy sighed with relief and told Mitch that Doug had had the flu over the weekend and she had asked him to stay home and rest.

"I'm sure he's just tired," Mitch said. "I'll send him back home when I see him."

Mitch never saw him again that day. He called Cindy two hours later, his voice more tense than before. This time, Cindy didn't take any chances. She explained to her boss that she had a family emergency and needed the remainder of the day off, and left work in a frenzy. As she drove away, she made calls to Doug's mother and sister and asked them not to mention anything to April, not until they found Doug.

Cindy estimates that there were more than fifteen cars circling a fifty-mile radius looking for Doug's car. They scanned road construction areas for possible places Doug may have crashed, they drove in and out of dozens of neighborhoods and strip malls, and they stayed in touch with all the emergency rooms. As the sun began to set, their tensions began to rise. April eventually came home to find a dozen family members convened in her living room, with only her father missing. Her face was perched on a cry when she asked what had happened.

The Beardens turned where they always turned in times of crisis: to the church, to Jesus, and to the Bible. Parishioners started a chain of prayer vigils while deacons drove the night streets trusting God would divine Doug's location. Hundreds of miles away in Ohio, Doug's younger brother, Tim, summoned the powers of his own congregation and encouraged the Atlanta-based Beardens with affirming scriptures. Finally, at five in the morning, they decided to enlist the help of the Atlanta Police Department.

The missing person's report had been active for only three hours when Cindy's phone rang. A police officer in Adel, Georgia, found Doug sleeping underneath a bush on the grounds of a plastics manufacturing plant. He explained that Doug's car had been found stuck in the mud of someone's backyard about a mile away, and he wanted to know if Doug had any problems with drinking. The officer's sleuthwork sounded naïve and endearing to Cindy's ear, but no matter, Doug was alive and in good hands, and a sigh of relief spread across everyone's face.

"He's just getting over a bad fever," Cindy told the policeman. "Just take him to a hospital and we'll get him from there."

Cindy and Doug's sister, Vicky, decided they would make the three-hour trek to Adel together, while April and her boyfriend took a separate car in order to retrieve Doug's Mustang. At just fourteen months younger than Doug, Vicky has a toughness that radiates through her strong back and resolved voice. Nobody even questioned whether they could keep her away from her brother, not at the time he needed her most.

After a little help from a towing company, April found that her father's car was strewn with gas station receipts and candy bar wrappers. Those bits of trash remain the only clues as to what actually transpired in the twenty-four hours he was missing. Even Doug has no recollection of that day; it was already lost, along with other large chunks of Doug's past.

When Cindy and Vicky arrived at Doug's bedside at Memorial Hospital of Adel, they were heartbroken. His arms and legs were pocked with irritating chigger bites, but Doug lay still, his eyes fixed on the ceiling. He didn't turn his head when Cindy and Vicky called his name.

"Honey, do you know who I am?" asked Cindy.

Doug looked at Cindy and his eyebrows furrowed. He paused uncomfortably and then a big smile erupted on his face.

"Of course I know my own wife," Doug said. "Have a seat."

Cindy slapped his shoulder and Vicky threatened to break his legs next time he pulled a joke like that. They sat down and immediately launched into their story about trying to find him. Doug apologized in a burst of tears and explained that he had no memory of the day at all, he was just sorry that he had frightened them. In the midst of his tears, the attending physician walked into the room. He told them that Doug was dehydrated and feverish, and that he personally suspected a form of meningitis—an inflammation of the layers surrounding the brain and spinal cord. They had already taken a spinal tap, but the result wouldn't be ready for three days—an unsafe amount of time, given Doug's symptoms. The only choice, he said, was to transfer him to Valdosta Hospital, where an infectious disease specialist would be on hand to offer consultation.

"They did everything to him at Valdosta," Cindy said. "CT scans, MRIs, steroids, acyclovir, you name it."

The neurologist took one look at the MRI scans and gave Cindy and Vicky her assessment. "He's got herpes encephalitis," she said. "We have to wait on the spinal tap results to be certain, but I'm ninety-nine percent sure."

Cindy and Vicky didn't hear a word after "herpes."

"It isn't that kind of herpes." She smiled. "It's fever-blister herpes, only it went to Doug's brain. He's got a really bad infection up there and it's causing a lot of cranial pressure. We're going to keep an eye on him to see if we need to place him on a monitor. He's not going anywhere for a few days."

The most frightening thing about the herpes virus is that it seems to know what it is doing. At any given point, HSV-1 traverses various parts of the body through the bloodstream, like a passive observer on a bus tour. It may occasionally choose a lip to blister, but in most cases it enters back into the bloodstream, disinterested in whichever organ it passes through. There is something about the moist, warm climes in the cranium, however, that coaxes HSV-1 into a small nerve pocket unprivileged by the immune system, the trigeminal ganglion. The virus then hovers suspended in the head indefinitely, until it mysteriously jumps the blood-brain barrier. HSV-1 then uses cranial nerves like a highway and most of the time targets the temporal lobes, as though it actually has a preference for them over other parts of the brain.

If left untreated, herpes encephalitis will kill 70 percent of the time; treated, it will still claim one out of four lives. Just as in Doug's case, the infected person shows flulike symptoms at first, which may or may not be accompanied by bizarre behavior. As the virus replicates within the cranium, it infects and destroys cerebral tissue along the way, causing portions of the brain to swell in retaliation. The swelling itself causes additional damage by compressing brain tissue, cutting off the flow of blood in affected areas. In survivors, HSV-1 infections cause motor retardation, seizure disorders, permanent memory damage, and higher-level cognitive distortions and impairments. In the most severe and tragic cases, the blood-choked portions of the brain will necrotize, so that upon craniectomy, the brain appears sickly gray and blotted with thick glistening globules of dead black tissue. Death from herpes encephalitis is slow, painful, vicious and, depending on the area it attacks, real or hallucinatory.

On the off chance it wasn't HSV-1, the hospital quarantined Doug's room, allowing only Cindy and Vicky access, since they had already been exposed to him. To keep his fever down, they placed ice packs under his arms. Within a matter of hours, Doug began hallucinating. Several times, he stood up on his bed and reached for the television, insisting that he needed to change the toll-free number on the screen. He began to tell Vicky stories about how he planned to blow up his house—stories so bizarre that they simultaneously amused and worried her. She sat and watched her brother slide down into an abyss of psychosis, and there was nothing she could do.

On the third day at Valdosta, Cindy heard the phrase "swelling of the brain" for the first time. With the doctor's cursory inflection, it sounded as mild as a sore throat.

"Don't worry," the doctor told them, "we'll get this bug out of his system. Doug has some swelling of the brain, but he's going to be just fine."

Neither Vicky nor Cindy questioned the meaning of "just fine." They took the doctor at her word and expected that once the antibiotic cleared away the infection, Doug would be back on his feet, just like before. They thanked the doctor and asked her if they could get him closer to home, and she smiled and wrote an order for Doug to be transported to Crawford Long Hospital in Atlanta later that same day.

Once he arrived at Crawford Long, Doug's new attending physician placed him in ICU, a shock and relief to Cindy and Vicky. ICU meant that he was in terrible shape, but it also meant that somebody would look after him around the clock. It also meant that his visiting hours were limited, even to family members, so Cindy and Vicky were finally relieved of their vigil. Each went back to her own home, showered in her own bathroom, put on fresh clothes, and rested in her own bed. During a conversation at home, Cindy and April decided that April would not want to see her father yet, not in ICU, not like that. She would stay home and keep the house in order while her mother tended her father.

While in ICU, Doug received visits from neurologists, neurosurgeons, and an order of medical students. Cindy began to get a sense that this wasn't a usual case, so she probed the visitors with questions every chance she had. Would Doug need brain surgery? Is his brain still swollen? How long until he can talk to us again? When can we take him home?

Somewhere in the answers, Cindy heard her first hint that Doug might not return to normal. A resident hoping to specialize in neurology sat with her on the couch and explained that the brain had swelled in response to the viral infection, and that treatment would involve a steady monitoring of his cranial pressure. The skull is already a crowded place, and once brain tissue starts pushing against the inner wall of bone, it can arbitrarily choke off critical blood supply. The resident then used a word that Cindy Bearden will never forget.

"Doug might have some *residuals*," she said, then echoed herself. "The virus does that, it leaves residuals."

"When a surgeon told us Doug may need a shunt, he didn't tell us what a shunt was," Cindy said. "When Doug stopped responding to our voices, nobody told us there could be damage. When his neck and the entire left side of his body stiffened, they told us it was just normal posturing. But when I heard her say 'residuals,' I felt absolutely lost."

Doug transferred out of ICU and into acute care almost two weeks after his disappearance. In the wide-open space of his new room, detached from sensors and feeding tubes, Doug looked like a ghost of his former self. His pale skin was loose and papery, the result of a rapid loss of weight—

over forty pounds in two weeks. The life that once beamed through his eyes now gathered in dark sunken pools on his face. He was talking, but just barely.

"Doug," Vicky asked, "do you know what day it is?"

"Happy birthday, sis," Doug croaked.

Tears rolled down her cheek. There was enough left of her big brother after all.

After weeks in an acute care setting, Doug progressed from a minimally conscious state to a minimally active state. He could speak, but often the sentences came out jumbled and unrelated. He had lost his continence somewhere in the process, and when the physical therapist attempted to walk him using a gait belt with two-person assist, he dragged the left side of his body along. In Cindy's eyes, her husband needed months of acute care; the hospital disagreed. After two weeks of minor progress, the social worker arranged for Doug's transportation to a post-acute rehabilitation center nearby.

The rehab didn't have a brain injury specialization, but Doug's personal drive compensated for their shortcomings. He began limping about the hallways without any help, but then began wandering off the floor, then outdoors. The rehab wasn't prepared to meet the special staffing needs of a severely brain-injured patient, so they put him in a Vail[4] bed. A cage of nylon netting atop a mattress, Vail beds look benign, like a miniature inflatable jumping gym. Patients can't choose to leave the confines of their mattress and can be let out only by an outsider. Many brain-injured patients report feeling humiliated, embarrassed, and traumatized by the experience of being locked in a Vail bed for weeks on end.

After three weeks in rehab, Doug was still disoriented, emaciated, incontinent, psychotic, unsteady, and hardly capable of clear speech or thought. According to the rehab, he was ready to move on. They called Cindy to explain that they had no more to offer Doug and to arrange for his transfer home.

"They notified me of his discharge, which I couldn't believe," Cindy said. "It was the first time I heard them mention insurance. They told me TRICARE wouldn't pay for any more treatment."

TRICARE is the military's tier-structured insurance. Despite more than twenty years of service, Doug qualified only for minimal coverage, called a 30 percent service connection. Although Atlanta has a reputable brain injury rehab, it wasn't an option for Doug. The only hope was entering him into the VA system, and the nearest place that could treat him was in Tampa, Florida.

"It's funny, because back then, I thought that sending him all the way to Tampa was a ridiculous option," Cindy told me. "Now, after hearing what other people have gone through, I feel lucky that we lived so close that we could still visit on the weekends."

Doug received better care at the Tampa VA than he ever had, but TRICARE gave him a maximum of six weeks, even after all the pleading phone calls Cindy made. His therapists gave him range-of-motion exercises. They put a wireless tracking bracelet on his foot and let him meander to his heart's content, and they gave him a battery of neuropsychological evaluations. By the end of his stay, Doug was walking slowly, he was continent once more, and he could take care of the activities of daily living, like grooming and eating. He recuperated physically, but now the extent of his brain damage was apparent. His short-term memory was shot, and his long-term memory fragmented. He was oriented to his person, but he was never quite sure of much else. The neuropsychologist gave Cindy a task list that included labeling every room and object in the house. He told her to get an alarm for all the doors and to avoid taking Doug to any crowded places. As he reviewed the list with her, Cindy couldn't quite believe what she was hearing. All this time, she had been expecting Doug to return to normal—sure, with "residuals," but not like this.

"Doug is probably as good as he's going to get," he told her. "Don't expect much more." Containing her tears, Cindy folded the list in her purse, said her goodbyes and thank-yous to the treatment team, and escorted Doug to the car.

In a matter of months, Doug Bearden went from a model of self-reliance to requiring round-the-clock care. He simply could not be left alone. His homebound lifestyle taxed him cognitively and psychologically. At times, he would march to his closet, grab his coat, and declare that he was going home when, in fact, he was standing in his own living room. In the middle of the night, he would roll over and tell Cindy that she wasn't his wife. At first she tried to reason with him.

"I am your wife, Doug," she pleaded. She took out the marriage certificate and showed it to him. "See, we both signed it. We're married."

"It's a fake. This isn't right," he said. "You're not my wife."

Once, Doug's mother took him to the grocery store and Doug started making small talk with her, as though he didn't know her well.

"After a while, I figured out that he thought I was someone else," says Jeanne. "It's just awful to realize that your own son might not recognize you."

At Christmastime, Tim Bearden paid his first visit to his brother. All the women in the family warned him beforehand that Doug wasn't quite the same, and they even sent him pictures to help him prepare for the encounter. As the baby of the family, Tim had always looked up to Doug as an inspiration, someone who challenged him at every turn. Over the past year, they threw around the idea of starting a church together, but the conversation was cut short by Doug's illness. Tim's first sight of Doug buried any hope of that plan.

"It wasn't Doug," says Tim. He has the same Bearden backbone as his big sister and casts a stalwart, compact figure to Doug's lean frame. "He looked different. He still had a limp and he was still confused. It was terrible for me to see him like that. I felt like I lost him, like he was there but he wasn't really there. The love wasn't different, but he was."

The encounter devastated Tim, but it also affirmed his intuition. He needed to be back with the family. Over the holidays, he talked with his wife, Beverly, about returning to Georgia. She had already served as a makeshift social worker for Cindy, but now she, too, sensed it was time to return home. It would take a great deal of work and heartbreak to disentangle their lives from the church and their kids from the community, but family is family. There really wasn't a decision to make.

By January, Cindy had used all her time allotted in the Family and Medical Leave Act, all her vacation time, and all the graces her boss could afford. She had no choice but to return to work. In an act of mercy, Vicky took an early retirement in order to help care for her brother. Even with Cindy's income and Doug's social security, the money coming in couldn't support a family of three, and Cindy's financial burdens began to mount. Perhaps in response to the stress he sensed, Doug's sleep became erratic, then minimal. His agitation began to simmer.

While lunching at Vicky's house one day, Cindy handed a bottle of pills to Doug and reminded him it was time for medication.

"I'm not taking that!" he yelled, and threw the bottle back at her. His outburst caused a moment of awkwardness but was otherwise ignored. Doug had never so much as raised his voice to her in the past, so she made excuses in her head about his having a bad day. In retrospect, she regards it as the first sign of the nightmare to follow.

Doug grew increasingly irritable. When he thought nobody was watching, his eyes would dart back and forth across the room. Cindy would occasionally hear him whimpering in the bathroom. Sometimes, he would let slip the delusions in his mind, yelling out, "Make it stop, make it stop!"

When his tirades and screaming grew incessant, Cindy felt she had no other choice than to admit him into Anchor Hospital, a facility for the acutely mentally ill. The personnel there claimed only a few experiences working with the brain injured, so they placed him in a room at the far end of the hall. The room amounted to little more than a holding tank. Nobody reminded Doug to brush his teeth. Nobody told him when it was time to change his clothes. And nobody was there to catch him when, from heavy sedation, he fell flat on his face and busted his chin open. There, the psychiatrist diagnosed him with the only label his *DSM-IV* would allow: personality change due to herpes encephalitis.[5]

. . .

Doug spent the summer pacing through the house at all hours. When his energy was spent, he would lay his head on Cindy's lap and tell her he was under attack by Satan. His clarity waned and vanished altogether, and then it seemed as though he was trapped in a terrible darkness.

"Am I dead?" he asked Cindy. He sounded serious.

"No, you're not dead," she replied.

"Are you sure?"

"I'm sure."

"When did I die?"

"Doug? What's the matter?"

"I'm dead."

"You aren't! You're—"

"I'm dead! I'm dead! I died!" he yelled, and stomped away. The deadness settled into his body. Where he once had a fluidity to his movements, he now had a robotic posture and gait. At the start of the year, he'd worn a long-suffering smile, but now his face looked hollowed out and expressionless. If his face showed anything, it was an occasional grimace of anger or a weary moment of fear.

"He started calling us bitches and whores," recalls his sister-in-law Beverly. "Then he would tell me he was dead, and he'd lay on the floor and not get up."

Most of the women in the family had a learned, built-in resilience to Doug's outrageous behavior, a blessing not bestowed on April. Watching her father deteriorate was too much for her. She stopped coming home most nights, calling to say that she was staying with friends. By the end of the year, she had spent more nights away than at home.

The Beardens concluded that Doug's rehabilitation in a brain injury program had been too short, so they decided they would do whatever it took to get him into one of the few rehabs in the state. After an initial evaluation, Restore

Neurobehavioral Center extended the offer of a bed but told the Beardens that they could not accept TRICARE. The only option was cash, and the best rate they could offer was fifty-five hundred dollars a week. At a time when Cindy was behind in house payments, the amount was crushing. Doug required several months' worth of care, so the Beardens turned once again to the church.

While someone attended Doug at home, other family members spoke during sermons and Sunday school. They held bake sales, yard sales, dinner benefits, and car washes. Their community of believers gifted them with almost seventy thousand dollars over a three-month period—enough for two months' worth of treatment. They paid the rehabilitation fee in cash and admitted Doug at the end of 2004. It wouldn't be enough time, the director informed them, but it would be something.

Doug's death delusion remained fixed but less insistent. He began to perseverate on food, constantly asking about the next meal, even in the midst of his current meal. He agreed to participate in therapies only with the promise of a snack—an obvious dilemma for the rehab nutritionist, who must keep patients on a consistent caloric intake diet. The neurologist, neuropsychiatrist, and internal medicine doctor continued to consult one another over the successes of Doug's medications and therapies. Just when Doug was evidencing signs of progress, his length of stay ended. There was no more cash, and the rehab discharged Doug with apologies and regrets.

The bizarre and erratic behavior continued at home. His food obsession wasn't just a passing compulsion; Doug's brain had a serious dysregulation problem, hyperphagia, the sign of a damaged hypothalamus. His brain kept sending ravenous messages, no matter how much food he stuffed into his stomach. If a person with hyperphagia is left unattended, they risk lethal injury by gorging. Extreme cases of uncontrollable hyperphagia can result in autocannibalism, the devouring of one's own flesh.

Cindy had a serviceman install a lock on the refrigerator. It didn't stop the incessant begging for food, but it did prevent Doug from gorging—with one exception. In a fit of exhaustion, Cindy opened up the fridge, knowing that it contained only cabbage and onions.

"You want to eat?" she said. "Go ahead, help yourself."

Doug tore into the cabbage and onions raw, as though he hadn't seen a meal in weeks. He choked down the entire head of cabbage, along with three large onions, skin and all, and went back to begging for food.

At sundown, he would have yelling fits, where he would scream out that his family had killed him. The next morning, Cindy trudged to the neighbors and explained that if they heard Doug yelling, they should ignore it, but if they heard her or April yelling, they should call for the police. After a few sleepless nights, Cindy stopped sleeping in the same room as Doug, preferring the couch or April's room instead.

The feeling of negation and loss seeped into Cindy's life as well. Caretaking exacted an enormous emotional toll on her. An entry from her journal during this time period reads as follows:

> I hate my life right now. I don't like this stranger that I'm living with now. Even Buddy [their dog] sees that he's different. God must hate me to put such a huge burden on me and I can't do a thing about it. It doesn't seem like Doug gives a damn about anything now. I have been

> watching him 24/7 for the past five months now, which is too much for even a married person. I feel like I have no life—no friends—I can't even run to the store alone. I've lost Doug and I've lost myself. I'm caught in this trap called Brain Injury Caregiver. Where do I go from here? No real husband, no friends, no job, just a caregiver to someone who doesn't know how to appreciate me. He can't even tell me he loves me. He can't even smile when he wants to. How can I survive this? What's to become of us? Will he ever be whole again? Will I always have to fear him? Will we ever make love again? What if my life is going to be this way for the rest of my life? Please God, I need more help. I'm so lost in this injury. I'm tired of reading about it, hearing about it, and living in it.

Late one night, when April was staying home alone with her father, she heard him raging near the refrigerator. He was trying to pick the lock with a kitchen knife, and with the passing moments he grew more and more agitated. When April attempted to dissuade her father from a possible binge, he locked his eyes on her and snarled.

"That wasn't my dad," says April. "It was the devil."

He drew the kitchen knife up against her and April shrieked, broke free, and ran around the house, her father chasing her and yelling. She ducked into her room, locked the door, and listened to her father pound and stab at the door while she cowered in a corner. She called her uncle, who appeared at the house just as Doug had managed to push April against a wall. Tim grabbed April and together they jumped into Tim's car and locked the doors. As they pulled away, Doug jumped onto the vehicle and started pounding at the windshield with his fist. The incident served as a vivid reality check: Doug Bearden was no longer safe to be around.

Tim moved in, but only long enough to get on the phone and start calling every facility in the state. After dozens of calls that ended in Doug's rejection on the basis of funding or no bed availability, the Beardens felt they reached a fork in the road. They had one option, and they had to risk it. Tim and Cindy drove Doug to the nearest VA hospital and escorted Doug to the admitting room.

Tim looked the admitting nurse in the eyes.

"You're going to take my brother," he said. "He's not coming home with us."

She started to explain how they weren't able to take people just because they were difficult, or—

"I'm dead," said Doug. "I'm dead, they killed me."

The nurse looked at Doug, disappeared behind a wall and reappeared with a psychiatrist. The VA told the Beardens that they would keep Doug for seven days, and not a day longer.

"We were determined not to answer the phone, in fear that they would let Doug go," Tim says. "They ended up keeping him for two weeks."

Instead of being an end point in Doug's treatment, the VA actually launched a series of facility placements. Within a six-month time period, the state healthcare system bounced Doug around like a hot potato, landing him in twelve different placements, all of them highly inappropriate for a brain injury survivor. When he was transferred to an Alzheimer's unit, the Beardens found him so sedated that he actually had food in his mouth from the previous nursing home.

Each facility had its own medication agenda for Doug, but they never gave him enough time to respond to the changes.

When he wasn't sedated, he alternated between incontinence, delusional states, psychosis, and hyperphagia. Now his health was failing. Even though Doug was no longer in the house, the Beardens refused to give up on him. They held weekly meetings, with half the time spent praying for Doug, and the other half strategizing treatment options. Although they remained hopeful, a year of hardship had also turned them into realists. Collectively, they decided that Doug's brain had experienced enough trauma. They chose to activate a Do Not Resuscitate order.

Doug's erratic gains and losses drove the Beardens to desperation. Their family meetings became more efficient and serious, as though they had established their own private case-management division, with several family members spending all of their spare time working on Doug's behalf. Beverly Bearden, it was decided, would tackle Doug's problem from a political angle. Day after day, she worked the Gordian tangle of Georgia Medicaid and TRICARE through a barrage of phone calls and e-mail. She contacted every state official, from the county on up to the governor's office, and kept contacting them. She explained that it was a matter of national embarrassment that the military wasn't providing the right care for one of its devoted servicemen, and she had all the paperwork to prove it.

Although it took months of meticulously logged communications, the strategy worked. The governor's office responded with a mandate to the VA, and the VA teased through all of Doug's medical history for a clue into the origin of his injury. They eventually found something on a single slip of paper from an outpatient visit with a military physician in 1981. A small notation read: "HSV-1, 2cm, Betadine ointment, no shaving 10 days." The VA issued Cindy Bearden a letter stating that they had officially determined that Doug's first indication of HSV-1 appeared while he was on active duty. He now qualified for 100 percent service connection, which opened the doors for total and complete care whenever he required it, along with additional monthly payments to cover the cost of home care.

In ways we may never fully appreciate, Doug Bearden lost his life. Within a period of three years, he endured over twenty different medical placements. His family was forced into bankruptcy, and every person he loves bears some type of scar, emotional or otherwise, from the experience. Cindy doesn't think she'll ever be able to enter the workforce again. Tim and Beverly are quick to thank God, but remain equally quick to act in case of a recurrence. Jeanne and Vicky drop by often to take Doug out shopping, or to play games, and the break is a cherished respite for Cindy. April still bites her nails and doesn't know what kind of future to expect with her dad. He hasn't had a death spell in ninety days, knock on wood.

Without rebirth and resurrections, humanity loses its heroes and loses its capacity for transformation. In order to gain life, the monomythic lesson goes, we must first lose it. In losing his past life of independence, Doug Bearden is enjoying a new resurrection. He told his wife that while he was dead, he couldn't feel anything. Now, life is full of feeling. It feels unsure and hesitant. It feels like a comfortable pair of blue jeans and a large recliner. It also feels like a locked refrigerator door and like getting teased by your daughter because you make sandwiches too big for your bite. Life feels like a tremendous sense of happening and not knowing. Life feels like this very moment, and then there's death.

ULTRAVIOLENT BRYAN

Bryan wouldn't answer. Kathy Larson had just unwrapped a tray full of tacos for her two boys. Christopher, her older son by two years, tore into his meal, hardly noticing the bizarre posture his brother assumed. Bryan, then five, was sitting upright, his head freakishly twisted and locked to the left, and his eyes fixed in space.

"Bryan, knock it off. Eat your food," Kathy told her son. Bryan didn't budge.

"Bryan? . . . Bryan?"

Kathy stared at her son, trying to piece together information she had heard from her experience as a paraeducator. Was this an attention deficit? Hearing problems? She called to Bryan again, but he still wouldn't answer.

"Bryan!" she yelled. Still nothing. Several more seconds passed before Bryan snapped out of it and started back at his meal as though nothing had happened.

In raising two young boys, Kathy let a number of bumps and bruises pass with a mere kiss and a hug, but this was something different. She knew it in her gut, but she wanted another opinion. She immediately called her husband, Bill, and told him about the strange spell. Bill said his boss had a son who had autism, but this didn't sound anything like the stories he had heard. Bill agreed that she should take him immediately into the urgent-care clinic and have him looked at. Why take chances, really?

"While we were there, he had another seizure," Kathy tells me. "They saw it happen, and they just said that we should take him to our family doctor. They didn't run any tests or anything."

I glance across the table at Bryan, who is thirteen now. He's quietly eating a slice of pizza and looks like any other teenager, but not like any thirteen-year-old. He's already six feet tall and closing in on two hundred pounds, most of it muscle. Bryan's countenance is so mild that he doesn't look threatening, just kind and content, an expression he borrows from his mother. It's difficult to imagine that only a few years ago, he was arguably the most violent child in America.

The Larsons live in a sunny suburban neighborhood, where each house resembles the one next to it. Everybody's lawn is crisply manicured, kids are zipping along on skateboards and dirt bikes, and at least one neighbor has a basketball net mounted nearby. You get the feeling that everyone here knows one another, that they're all active in their neighborhood association, and that they invite one another over for cookouts. Horsethief Canyon Ranch is a safe place, usually.

Bill Larson has white hair but radiates a boyish manner. He's barrel-chested, with dense, sizable forearms, and he has a surprisingly gentle handshake. Bill works for a large construction company as a trainer and developer. He introduces me to Kathy, who in her blue jeans and comfortable sweatshirt looks like the mother of two boys. Her face is tolerant and mild, and her voice is sweet and relaxed. I try to picture Kathy screaming at the top of her lungs for help and Bill pinning his son to the ground, but I can't piece together the image. Kathy invites me to sit down at their dinner table and join them for some pizza.

Bill and Kathy were living in this same two-story house when they were told that Bryan had a seizure disorder, a woefully incomplete diagnosis. The seizure disorder came as a real shock, as nobody in the family had a history of epilepsy. After that first spell in February, Bryan evidenced two or three seizures a day, and it was starting to affect his work in kindergarten. Their pediatrician referred them to a neurologist, who took readings from an EEG and recommended a course of anticonvulsives, and the Larsons nodded their heads and agreed that the seizures should be controlled, and if possible, stopped.

The medications had an effect on Bryan, but not on his seizure activity.

"On Fridays I would be the parent assistant in Bryan's class," Bill told me. "With the medicine, Bryan would go into a 'whatever' mode. He was a lot slower, and he started having problems using scissors. Instead of being one of the first ones done, he was one of the last. It was an immediate contrast."

Subsequent trips to the neurologist left Bryan on a medley of serious medications, but the absence seizures persisted. Finally, after several months of varying dosages, the doctor returned to Bryan's initial prescription, and it actually seemed to work.

"His seizures were finally under control," Kathy said. "But that's when his behavioral problems started."

Just after turning six, Bryan developed a new attitude of defiance that shocked his parents and teachers.

"At first, he refused to do his work in class," Kathy says. "Then he started refusing to come in after recess. He became oppositional and the doctors simply blamed his behavior on the medication."

"We assumed the doctors knew what they were talking about," Bill commented. "We thought it was no big deal, and we thought Bryan would outgrow it. But then his seizures started back up, his behaviors worsened, and he started holding his one arm to the side a lot. We took him back to the neurologist, and it wasn't until June of that year that they finally ordered the first MRI of Bryan's brain."

The MRI showed an odd formation on Bryan's right temporal lobe, which the technician and the doctor both perceived as scar tissue. They theorized that the lesion may have been the result of a high fever in the past, and that with time, Bryan's brain would simply heal itself. Until that day came, it was best to keep Bryan's seizures under medicinal control.

"Nothing was working," Bill said. "Nothing."

At school, Bryan's refusals turned concrete and his oppositions nonnegotiable. Both Bill and Kathy regularly fielded phone calls from frustrated teachers and the principal. The school trusted that Bryan's unruly behaviors were rooted in his medication, but the simple reality was that there were other children in the classroom who didn't deserve the disruptions. A solution needed to be found quickly. By the time November came around, Bryan's behaviors were caustic but still manageable. The situation inside Bryan's skull, however, was another matter. When a friend reported to Bill that Bryan's left arm remained dropped even while on the swing, Bill marched his son back to the neurologist's office and demanded another MRI.

"Absolutely not," the doctor told him.

"Please," Bill begged. "I will pay for it out of my own pocket if I have to."

The neurologist continued to argue with Bill, but Bill's stubborn insistence eventually wore the doctor down. He flipped through Bryan's records once more, and then paused. He thought for a moment and looked up at Bill.

"Okay, let's try an MRI again, but with contrast this time," he said.

Although he had undergone an MRI scan before, Bryan felt frightened this particular time. He responded with his only defense: rage. He started swinging and kicking at the staff and ended up requiring a chemical sedation just so he could be placed into the scanner. The technicians hurried through the procedure, and the Larsons whisked Bryan home before he threw another fit. As soon as the sedation began to wear off, however, Bryan's mood only worsened. He started ranting and stomping throughout the house, kicking at the walls and yelling.

"I'm going outside!" he screamed. "I'm going outside and I'm running into the street so I can get run over by a car!"

"He was only six when he said that," Kathy told me. "I was completely scared and freaking out inside."

Only one thing frightened Kathy more. Shortly after the contrast MRI results came back, she received a message from their doctor saying that he wanted to see the Larsons first thing Monday morning.

"I was in denial, the news was so bad," Kathy says, shaking her head.

The neurologist told the Larsons that their son had a brain tumor, and that, thank god, it was not the spider type, but the enclosed type. The contrast showed a white growth, an astrocytoma, on Bryan's right temporal lobe, the same location as the alleged scar.

"They're doing great things now with tumors," the doctor said, "but it's still serious. You have an appointment this afternoon with a neurosurgeon. This guy is going to want to remove it as soon as possible."

Although the news was terrible, Bill and Kathy felt a sense of relief. At last someone could explain what was wrong with Bryan, and there seemed to be a clear solution.

"We thought that we were going to go to the hospital, they would take out the tumor, and everything would be fine again," Bill said. "That's the impression we got."

Fine or not, the thought of her son on a brain surgeon's table was enough to crumple Kathy into tears. She cried when she showed up to gather Bryan's belongings at his school, and the educators there expressed shock and sorrow at the diagnosis. How could they have missed a tumor, they asked her. How big, how long, how fast, where at, what type? Kathy could reply only with puzzled glances and nods. She knew little more than they did.

Later that same afternoon, Bill, Kathy, and Bryan found themselves sitting in a surgeon's office seventy miles away, watching an instructional video about brain tumors and the procedures involved in their extraction. For all of our technological advances, neurosurgery still amounts to brain cutting. Most brain surgery is simply the creation of a new injury to avoid the potential effects of a more significant injury. In Bryan's case, the surgeon planned on opening a quarter-size hole near Bryan's temple and working out the tumor with the same suction hose we encounter during dental checkups. After the video, the surgeon explained that he expected to collect about 10 to 15 percent of the surrounding tissue, a passable amount of lobe that nobody, including Bryan, ought to notice.

Bryan's first surgery occurred shortly after Thanksgiving Day, a full ten months after Kathy first reported his seizures to an urgent-care doctor. Because he was so wild before the surgery, Kathy and Bill never had a chance to hug and kiss their son—he was sedated and immediately whisked away to the operating room. The surgery took a total of six hours, but the wait would be worth it. The Larsons hoped that Bryan's seizures and his outbursts would be extracted along with the tumor. They longed to have the old Bryan back.

"When he came out of the surgery, he was all wrapped up, swollen, and had purple eyes," Kathy recalls. "You were almost afraid to touch him."

In the postoperative consultation, the neurosurgeon beamed proudly and announced that he had removed the entire tumor. No more seizures, no more erratic behavior, problem solved. Bill and Kathy shared a deep sigh; they had complete trust in the surgeon. The doctor cautioned the Larsons to treat Bryan's brain carefully for the following year—no roller coasters, no contact sports, no roughhousing. Bryan began to emerge from his sedation shortly after the conversation, and within two days he was back home, happily helping his mother decorate the house for Christmas.

Scarcely a week after the surgery, Kathy noticed the change.

"Bryan," she said. "Lunch is ready."

"Bryan? . . . Bryan?"

Bryan was frozen, his consciousness disenfranchised from his body. The seizures had returned, only this time they served a different purpose. They were a warning sign, a hint at the fury about to follow.

"It makes me want to cry just thinking about those days," Bill says.

I haven't known Marty McMorrow long enough to know whether his subdued and easy demeanor is natural or the result of the twenty-five years he's spent around violent and aggressive brain injury patients. Either way, in his presence, your shoulders tend to relax a little and you feel a general sense of openness, as though the room just doubled in size. In the nineties McMorrow, a behavior analyst by training, designed a noninvasive neurobehavioral program adopted by over a dozen brain injury rehab centers in America, but this humanistic approach is largely unknown among mental health practitioners. At a hotel in downtown Baltimore, I had the chance to catch up with Marty, and I asked him if he was optimistic about the outlook for brain injury survivors who suffered from behavioral problems.

"Optimism is too strong a word," he said. "There's too much social resistance, and there's too much individual resistance. A lot of people have trouble dealing with aggression, and that's just human nature.

"When a person swings at you in the face, you're going to get mad and things can get really out of hand," he explained. "But if you break down the behavior, what is really your most natural response? It's surprise, and maybe shock. It isn't anger. Many of us who work in human services have lost our natural human reaction."

Marty's experience among the brain injured is peppered with stories of mistreatment at the hands of medical professionals. On one occasion, he was asked to evaluate a former model who had sustained a brain injury and was currently living on a psychiatric ward. Prior to meeting with the woman, he looked through her medical chart and learned that she had been placed in isolation for seventeen hours a day for the past thirty days in a row. He conducted the evaluation in a seclusion room that contained only a mattress—itself a passively hostile environment for the patient, as she had also lost her legs in the injury. After talking with her at length and scrutinizing the chart for details, McMorrow realized the woman had been secluded simply because she was annoying the staff with her excessive talking and anxiety. Once the woman received the appropriate care in a dignified environment, her behavior changed radically.

After visiting hundreds of patients in various facilities, Marty has witnessed numerous indignities heaped upon brain injury survivors under the aegis of restraint and seclusion. Yes, the patient is violent. Yes, they are agitated and aggressive. But they are human. They always retain that, their humanity.

"It's absurd what some people come up with when dealing with aggressive patients," says Marty. "You're not going to set limits with a six-foot-five, four-hundred-pound brain injury patient! Who are you kidding?"

Since the 1950s, behavioral modification has become an ingrained protocol in many U.S. psychiatric hospitals, and the accompanying mentality has remained largely militaristic. You set boundaries, you reward good behavior, you punish the bad, you take away all privileges, and you slowly return them. The patient will submit, the thinking goes. Extreme behavioral modification is rational, logical, and, in the treatment of brain injury survivors, reductive and inhumane.

Shortly after the holidays, Bryan returned to school and seemed enthusiastic about rejoining his class. He was coloring in the lines, he was obeying his teacher, he was getting along great with the other first-graders. The doctors had told the Larsons to give his brain a little more time to heal, that the absence seizures might just work themselves out. When the school day ended, however, a different Bryan seemed to show up.

"Anything would set him off," his brother, Christopher, told me. "I remember once we heard an ice cream truck, and so we asked our mother for some money, and she said no, because we were having dinner. Bryan didn't like that."

Bryan would tense up his fists and start screaming and yelling. He stomped around the house, furiously pounding his fists at the walls, threatening and raging against anything in his line of sight. His fury seemed only to heighten with the passing minutes. Eventually, Bill decided that he needed to grab hold of Bryan before he hurt himself or someone else. He hugged Bryan in his large arms and whispered to him to calm down, that he would be okay soon. Bryan squirmed and tossed for half an hour, and then his mood quieted. He panted a little and then went back to playing as though nothing had happened.

The fits continued with greater and greater frequency.

Bill asked his boss whether his autistic son had any behavioral problems, and his boss admitted that restraints were just part of taking care of his son. He explained that children could be physically held in a safe manner until the outburst passed, and he even showed Bill a few moves that he used on his own son. Bill took the advice to heart and began using the techniques when Bryan flew into rages at home.

"I would grab him from behind with my arms around his torso," explains Bill. "Then I'd toss my legs over his legs, and we'd stay that way, sometimes for almost two hours."

"At first, we thought he had some control over his behavior," says Kathy. "The outbursts were only happening at home and not at school. But then he had an incident."

After recess one day, Bryan started fuming and pacing wildly around the classroom. His teacher demanded that he return to his seat, but he refused. Bryan started swiping his arms over the shelves, dumping out containers and knocking over stacks of paper and books. He tore at the posters on the wall. His eyes grew wild and maniacal and he began screaming. In his rage, he yelled that he had a gun and that he was going to shoot and kill everyone in the school, causing the classroom full of first-graders to break out in chaos and tears.

"School shootings were in the news at the time," says Kathy. "So the school reacted by suspending Bryan for ten days."

By March of that year, the doctors agreed that another MRI was in order, but this time, they said, it really did show only scar tissue. They would try to control Bryan's behavioral problems with medications, but outside of that, they offered the Larsons no certain hope of stopping the outbursts. Meanwhile, the intensity of Bryan's attacks continued to escalate.

"I was restraining him a couple of times a week by myself," says Kathy, "but Bryan would literally wear me out after an hour or so. I would yell for Chris, and he would run and get Andrew."

Andrew is the Larsons' neighbor, a police officer built like a lineman. He would rush in the front door and hold Bryan in place while Kathy disentangled her limbs from Bryan's. Bryan might rage and squirm for another hour, but once his fit ended, he would apologize to everyone and express genuine remorse. Over time, Andrew became one of Bryan's confidants—Andrew's size alone was enough to reassure Bryan. After all, Andrew could protect Bryan from himself.

"We never thought it was anything else other than his brain causing the rages," says Bill. "He wasn't even Bryan; he was like a caged animal."

The violent episodes became so frequent that Kathy had to quit her job and stay home with Bryan, who simply wasn't ready to return to school. Although she had the best of intentions, she soon found herself unable to contain Bryan during his more aggressive attacks. She began to call Bill home from work regularly.

"After a few weeks of restraining my son, my arms looked like they had been beaten with a baseball bat," Bill says. "He had bitten me so many times I had to get a tetanus shot."

On several occasions, Bryan tried to break loose from the hold by bending his father's thumbs back to their near-breaking point, causing damage that Bill complains about to this day. The appropriate restraint holds were simply becoming ineffective against Bryan's frenzied struggles. He would pinch, bite, pull hair, and spit. At one point, Bill realized he could no longer hold his son the same way without getting hurt, so he began pinning Bryan to the floor on his stomach.

"He was only fifty or sixty pounds at that point," Bill recalls. "But he was so violent that he would buck me around like a horse. Chris would try to stay on his legs, but he got kicked off every time. He would start cursing, too. Bryan would yell out words that you only learn after twenty years in the navy."

Once, while pinned belly-down, Bryan grew so crazed that his face turned red with rage and he bit at the carpet, gnashing and pulling at it until it ripped away from the floor.

"You had to hold on with all your might," Kathy says. "Every time he got away, I was afraid he was going to kill himself."

On a different occasion, Bryan overpowered his mother, ran upstairs, and attempted to dive out the bathroom window on the second floor of his house. She and Chris pulled him back inside before he made it. More than once he bolted out the front door, only to be found out of breath, blocks and blocks away.

Bill and Kathy thought that if Bryan was back in school, it might cause him to have fewer outbursts. So they reenrolled him, but this time with a different protocol. Bryan would have two aides accompanying him at all times. After a short trial period, the school opted for a preemptive approach. They put Bryan into his own classroom with his own teacher. The three adults would spend the bulk of their day either restraining Bryan or chasing him.

"My friends would tell me that they saw my brother sprinting down the hallways again," says Christopher. "It got to be just a normal thing."

"One day they actually took him down on the hot asphalt, in front of all the other kids," Kathy says. "It was horrible, just horrible."

At home, the Larsons were adapting to Bryan's attacks. They noticed that violence always followed a cluster of seizure activity.

"When you saw him start breathing heavily, you knew you had about one minute before things got really ugly," Bill says. "Kathy would check her pockets to make sure she had her phone, Chris learned Andrew's schedule so he could find him in a moment's notice, and I always had my pager with me."

The school called Bill and Kathy so frequently that they eventually learned to say "no." They felt that the school needed to assume some of the responsibility in caring for Bryan, otherwise nobody would understand the complications that Bryan's brain injury posed for them at home. Within a few weeks' time, the school system assigned a psychologist and a crisis team to analyze Bryan's school and home environment and offer any consultation they could. Unfortunately, the team was untrained in brain injury, and so they approached Bryan's behavioral modification as a mental health problem.

"One time he actually went into a rage while the crisis team was right here in my living room," Kathy says. "Bryan ran all around the house and nobody could catch him. He eventually got out the front door. We were right in the middle of another crisis, and the crisis team looked at me and told me that they needed to get to their next appointment. That's when I lost faith in the system."

Kathy Larson is not alone in her loss of faith; a majority of caregivers I speak with express the same disappointment. If it weren't for the true believers in brain injury rehabilitation, a lack of faith would be the least of a caregiver's worries.

. . .

"I've seen some amazing things happen when you treat people with dignity," Marty told me, recounting his experiences running a brain injury center. "Like the guy that killed this moose with a motorcycle."

Marty explained that several years prior, Zach Greenburg survived his moose collision only to require a significant resectioning of his frontal lobe. Hospitals and psychiatric units claimed they could no longer handle his extreme level of verbal and physical aggression. While brain injury rehabilitation should have been the first course of treatment for Zach, it came late, and the consequences were evident to Marty.

Zach was transferred to Marty's care in a Chinook helicopter. When Marty arrived at the airport, he saw Zach descend from the craft wearing full restraints, with his wrists and ankles chained to his belt. Because of Zach's constantly agitated state, nobody had ever tried to shave him since the injury, so his face was blanketed in a wild tangle of hair. The ruse didn't deter Marty.

"Go ahead and unchain him," Marty told the escorts, who didn't believe him at first.

Before Zach could lunge away, Marty established eye contact with him. He peered into Zach's distant gaze and spoke to him.

"Can you look at me?" Marty asked, and repeated, "Can you look at me?"

Zach's gaze focused on Marty's face.

"You're not going to need those restraints anymore," Marty told him, and led Zach into the terminal.

Immediately, Zach began orienting himself to his surroundings. He walked over to an airport chair and tried to pick it up. Marty, along with a brain injury tech, began implementing a technique called partitioning, where a patient is afforded extra space while still being assisted. Zach walked toward a janitor's closet and tried turning the locked doorknob. He then walked into the men's restroom, and for the first time since his injury, he toileted himself.

"For a while, he believed he lived on a submarine, and he ordered drinks from large plants," Marty said. "Most people assume that's a psychotic state, but Zach had a navy background. It turns out he was just incredibly confused."

Within a few months of brain injury rehabilitation, Zach couldn't believe that he had ever been a behavioral problem to anybody. He expressed shock and concern that people feared him, and he had virtually no memory of his aggression.

"We can so frequently make such dramatic impact," Marty said. "It may not always be the case, but in roughly sixty to seventy percent of cases, it does turn out that way."

•

Bryan's rages at school and at home demanded action. The Larsons felt they had no choice but to commit Bryan to a children's psychiatric unit. They accepted the doctor's explanation of brain tissue scarring and assumed that the only course of treatment available to Bryan was psychiatric care. Bill and Kathy signed the appropriate papers and entrusted Bryan to the hospital with the hope that his behavior might take a turn for the better.

"At that point, I thought he had no future," says Kathy. "I thought that he would be in a locked-down facility for the rest of his life."

The environmental change and the close supervision brought Bryan a small amount of relief. After seven days in a supervised medical environment without a major behavioral incident, Bryan evidenced no symptoms and his condition seemed harmless. The psychiatric staff began to doubt the Larsons' struggle. Maybe it was their fault, a therapist suggested. Maybe something was happening at home that was upsetting him. Bill and Kathy escorted Bryan out of the facility, but the moment Bryan hit the parking lot he flew into another rage. Bill had to restrain Bryan on the grounds of the hospital that had just discharged him. Both Bill and Kathy began to feel despondent about the decisions they had made at the encouragement of their doctors. Surely, they thought, someone must be able to help Bryan out.

While on their hunt for a new treatment program, the Larsons took Bryan to his one-year postsurgery checkup, where he received another contrast MRI. Once again, the Larsons were brought into the doctor's office, this time with news even worse than before. Bryan's right temporal lobe was speckled with a constellation of tiny astrocytomas—or so it appeared. Further scans exposed the actual, unanticipated suspect: a dysembryoplastic neuroepithelial tumor (DNET), a slow-progressing lesion with neuronal abnormalities—a characteristic that causes DNETs to skirt detection. Bryan's DNET was confined to the temporal lobe, but there was a strong likelihood that it might encroach on the hippocampus and amygdala—key components of the limbic system thought to play a role in memory encoding and emotional regulation. The Larsons' neurosurgeon told them that it was best to keep a close eye on the DNET, and that Bryan would need to undergo extensive testing before they could offer the only proper course of treatment: an extensive temporal resectioning, a nearly complete removal of his temporal lobe.

Life could not be suspended; it wasn't the Larson way. In the face of the crushing news, Kathy and Bill were determined to continue with Bryan's education. They began to research organizations that claimed to have experience with behaviorally challenging cases and discovered a school named Elm Park.

"The school was basically the last stop before prison for most of these kids," explains Kathy. "But their advertisements said they had a neurobehavioral program. We decided to give it a try."

With only one employee familiar with brain injury, the term "neurobehavioral program" was a stretch. Elm Park was full of juvenile delinquents who had either criminal records or emotional problems so severe that they were too disruptive in any other environment. At eight years old, Bryan was the youngest child on the campus. Bill Larson made numerous phone calls to Elm Park officials, trying to figure out if they were prepared for Bryan, and each call he made was met with reassurance. They would pick Bryan up personally for his first day of class.

"I followed them all the way to school that day," Bill recalls. "I wanted to see exactly what they planned, because I had a suspicion they had no idea what they were doing."

When Bryan entered the building, the security guards had him empty all his pockets and his backpack. They paged through his schoolbooks and they passed Bryan through a metal detector, then into a room where he received a full pat-down by another security guard. Bill walked with Bryan as he was escorted to the doorway of his classroom.

"Dad," Bryan said, "you can leave now, I'll be okay."

Bill smiled at his son, patted his head, and walked down the hallway. Before he even made it out the doors of the school, a teacher ran to Bill and exclaimed that nobody could find Bryan. He had disappeared as soon as Bill turned his back. Bill offered to help them look, but he was unfamiliar with the school. The staff members told him to go ahead and leave, that they would organize a search for Bryan. Bill shrugged his shoulders and made his way back to the car.

"I started driving along the front of school, and there's Bryan, tearing across the front lawn, being chased by several guys," Bill said. "I told myself that if this was the only kind of school available to Bryan, we were in big trouble."

While attending Elm Park, Bryan begged his parents to dye his hair black, and he asked for a black ski jacket for his birthday. He wanted to fit in, he said. Although the Larsons couldn't prove Bryan was being picked on, they sensed he was being mistreated. The school counselor accused Bryan of faking his seizures and using them as an opportunity to wreak havoc, and to an untrained eye, Bryan's attacks must have seemed malicious. He darted under parked cars, tore up his textbooks, and caused a daily stir in each of his classes. The chaos was not always his fault, however. During one attack, a classmate sitting behind Bryan set fire to his hair in the middle of class.

The state of California could offer no other realistic options for Bryan. A mental health professional visited the Larsons' home and offered to help on behalf of the school board. After he witnessed one of Bryan's lesser rages, he confessed that the Larsons' situation was far beyond his scope, and he couldn't come up with a suitable solution. Kathy's research turned up one school in Massachusetts that seemed as if it could offer some direction for Bryan, but the thought of sending him so far away from home felt as disastrous as not getting a proper education.

"We were fed up with Elm Park and we were fed up with our school district," Kathy says. "That's when we heard about a guy who was starting a school for children with brain injuries right here in Southern California. I called him up and he told me that they were looking for kids who had behavioral problems due to brain injuries. I thought that it was too good to be true."

News of the school was so premature that it felt untrue for the sixteen months that passed while the Larsons waited. As Bryan endured an endless battery of neurotesting and medication management, Bill and Kathy began their battle with the local school board. According to the U.S. Department of Education, Bryan's brain tumor and the ensuing trauma it caused to his brain don't exist. Brain tumors are classified as "other health impairments" and don't fit the federal definition of brain injury. The Larsons first needed to demonstrate that Bryan's brain damage did in fact exist, and that it required the same attention as all other brain injuries. The school board required Bill and Kathy to create and present an Individualized Education Plan (IEP), which would note Bryan's medical condition and environmental needs. The plan acted as a request for special services so that Bryan could achieve realistic goals in his education.

Now infused with a new sense of caution born of myriad letdowns, Bill Larson hired an attorney to help create their first IEP. After the first meeting with the school board resulted in a shocking dismissal of the Larsons' requests, the attorney informed Bill that it would cost three thousand dollars to continue on with his help. Bill and Kathy opted for a self-designed crash course in educational policy reform.

"We started taking classes and seminars and school district training sessions," explains Bill. "We learned how to do IEPs, how to file complaints, how to work with the office of civil rights. Many times it would be just us meeting with twenty-five people in the school district and their attorney, with just Kathy and I on the other side. And we would win."

An informal IEP might be a sentence or two. The first IEP Bill and Kathy signed was three pages long. The Larsons could never have anticipated the amount of paperwork, meetings, evaluations, and assessments they would endure as a result of championing Bryan's education and care. When I went through one of Bryan's functional analysis reports, which included a history of his parents' efforts, I counted over twenty-three separate reports filed, aside from the monthly IEPs that Bill and Kathy presented. By default, Kathy Larson became a full-time advocate for Bryan, and Bill spent all of his time outside of his job joining in the effort. They had long since abandoned hopes of regaining a normal social life. In the course of advocacy work, they came to learn of other brain-injured children in similar predicaments and soon understood that this crusade needed leaders. The Larsons weren't about to back down.

Bryan's third-grade year at Elm Park concluded with a withdrawal from school due to medical reasons. Another MRI late in the spring semester indicated that the tumor was progressing and that surgery needed to occur as soon as possible. The neurosurgeon explained that he would have to remove the bulk of Bryan's right temporal lobe, along with any other portions of the brain that the tumor had overtaken. Bryan might get better and he might not, the surgeon cautioned, but at least the tumor would be gone. Bill and Kathy were relieved to have an honest, straightforward answer. The surgery was set for May.

During the procedure, the tumor's reaches forced the neurosurgeon into a major temporal resectioning.[1] The deeper the neurosurgeon dug into Bryan's brain, the more he was forced to rely on his own vision and anatomical understanding. He shined bright light into the cavity to discern the DNET by its slight discoloration, and tested his suspicion by prodding the growth to determine if it felt different from surrounding healthy tissue. At the most trying part of the procedure, the neurosurgeon's hands pulled and seared away bits of tumor within millimeters of the brain stem, where the tissue that housed Bryan's most integral functions—breathing, swallowing, coordination—lay exposed. The surgery took all day, as the neurosurgeon had to painstakingly tease sections of the tumor out from a wet tangle of arteries and tissue.

Bryan emerged from the operating room with his head wrapped in gauze and with tubes protruding from his bandages. When he awoke less than ten hours later, the Larsons learned that his jaw was clamped shut, and one of his eyes wouldn't close. Later, the neurosurgeon explained that Bryan may have experienced minor nerve damage to his face, but that those kinks should heal up soon. Although Bill and Kathy had researched as much as possible, it was still difficult to guess what Bryan's postsurgery life would be like. They understood what Bryan was like with damaged brain tissue, but they had no frame of reference for what Bryan would be like with an entire lobe missing. They weren't quite sure what part of Bryan was gone, and what remained.

The tumor had long ago destroyed the portions of Bryan's brain that allowed him to regulate and contain his rages. With the critical tissue now completely removed, Bryan would still be susceptible to the same kinds of problems. His brain was now distressed and physiologically disposed to aggression and violence. Bryan would have to be taught how to use his brain to heal itself.

We tend to think of our brains as a coil of fixed connections, like a complicated electrical panel that reroutes messages back and forth. In recent years, neuroscience has revealed that the brain has a dynamic proclivity for self-recovery. In a global sense, the brain can actually relocate functions from one area of the brain to another. On a microscopic level, the neurons themselves can react to changes by making new connections. From a certain perspective, the brain acts like a switchboard that rewires itself when there's a short circuit. This sublime attribute, brain plasticity, yields wonder upon wonder.

In several studies of children who endured entire hemispherectomies (a procedure where half the brain is removed), a majority of the children demonstrated shocking results.[2] Language centers, which were long thought to be relegated to a specific hemisphere, reemerged in the remaining brain tissue. Sensorimotor skills that should have been missing were regained, so that survivors could run, play Ping-Pong, and pick up the piano; they blossomed into productive, independent adults. According to the fixed-model view of the brain, that isn't supposed to happen. The mystery of plasticity continues well into adulthood, allowing even elderly brains the capacity for restoration. Years after a brain injury occurs, adult survivors continue to make gains that surprise even the most optimistic doctors.

Prior to Bryan's surgery, Bill and Kathy learned that the Alex Center,[3] the long-hoped-for neurobehavioral school, would finally open that fall. The school district bent under the Larsons' methodical determination, and they agreed to fully fund Bryan's education at the center, as well as his ninety-minute commute to the school. The only transportation company willing to contract with the school board happened to be a limousine company. Each day of his fourth-grade year, a black stretch limo pulled up to the Alex Center to drop off and pick up lone Bryan Larson.

Sharon Grandinette, the center's then-director, remembers her first impression of Bryan clearly.

"He was a cute little blond-haired boy," she told me. "If you saw him on the street you would never know he had a disability."

As a special-education brain injury consultant, Sharon regularly presents talks at national conferences, and most of her topics center around the hardships that brain-injured children face when dealing with inadequacies in their school districts. After one of her talks at the North American Brain Injury Society in Miami, I met her for a beachside lunch. I asked her what Bryan's initial experience at the Alex Center was like.

"Bryan was excited on the first day," she recalled. "He said it didn't really look like a school. The facility was once a skilled nursing facility that was attached to a larger hospital. We had done a lot of work to convert it into a school. There were murals on the walls, we had wide hallways, and we had handrails. It didn't look like the kind of school Bryan was used to, but he knew it was better than where he was. He wanted very much to be a good student. He was great—for the first couple of hours at least."

The brain injury educators at the center had to teach Bryan how to learn again—reading, writing, and arithmetic could come later. It was more important to help Bryan regain the fundamentals: how to pay attention, how to socialize appropriately, how to constructively direct his anxieties, how to stop the rages. Through rigorous preadmission assessments, the teachers figured out that Bryan's triggers appeared to center around academic activities.

"You have to understand that Bryan was typically developing until his brain tumor came along," Sharon said. "He had a pretty good idea of what he was like before. He was writing and reading, he knew what he was supposed to be capable of at his age, and he knew that he couldn't do it. Knowing that, you have to be careful how you introduce the work."

Bryan began the day with a preferred activity, which for him always consisted of music. His teacher showed him a picture board that contained visual representations of the activities he was to accomplish each day, and explained how each time he completed an activity, he could rip the Velcro picture off the board and put it in the "done" pouch. There were several activities Bryan was supposed to perform by day's end, and every staff member knew that each one of them contained a potential trigger. They knew Bryan would go ballistic, they just didn't know when.

"We all went through training with a board-certified behavior analyst," Sharon said. "He taught us how to antecedently manage the environment. One of the holes we typically fall into when our buttons get pushed is the method we learned from our parents, which is 'Stop that, don't do that, go to your room.' All that does is trigger brain injury students."

That day, Bryan's first act of defiance began with refusal. He set his head down, and then a moment later he sat up and pitched his pencil across the room. He tore up his paper, and the teacher, reacting as her parents had probably done, looked dismayed. She scolded Bryan, saying that she had worked hard preparing the paper for him. Before she could finish her reproach, Bryan was out the door.

"You want to talk about a kid with athletic ability," Sharon said, "this kid was fast. He could run three or four laps around the entire hospital building at full speed."

After a short learning curve, the staff at the Alex Center realized that chasing Bryan was futile. Instead of provoking him, they learned to station themselves at various points around the building and let Bryan exhaust himself. Once he started panting, they would tell him that he could come inside when he was ready. On one occasion, however, the freedom backfired. In midsprint, Bryan caught sight of a drainpipe and scaled it with the agility of a seasoned rock climber. He was on the slippery clay tile roof in seconds.

"We tried everything to coax him down, and we called the police and fire department. He wasn't calming down and he moved around a lot. Just talking to a person on cognitive overload irritates them even more."

Eventually, a behavior analyst was able to lure Bryan down from the roof by telling him that he had already made significant progress that day, and that he still had time to accomplish activities.

The policemen, however, saw Bryan's behavior differently. They approached Sharon with the intent of placing Bryan under a 5150, the police code for the transport and detention of a severely mentally ill person, but Sharon refused. She went through great pains to explain the difference between brain injury and mental illness to the police officers, and she told them that Bryan's emotional state had become ungoverned by the lobectomy. The police reluctantly agreed to leave Bryan under the center's care once more.

Bryan's incident on the roof led to the creation of the center's quiet room, a separate room where the student felt in control. Students could decide how long they wished to stay in the room, and staff members were not permitted inside, nor were they allowed to coax students out. Bryan availed himself of the room's peaceful environment on more than one occasion. The quiet room helped, but Bryan was still susceptible to full-blown attacks.

"At one point, the running was not helping and we were afraid Bryan was going to hurt himself," Sharon explained. "We called the behavior team to grab the blue mats. Bryan knew what was coming. It took one staff on each limb, and we knew that once we were into this, we were committing for at least an hour and a half. Legally and ethically, after twenty minutes, you have to let go of a child. Well, the second we let go, he would try to smack us. He then bit his tongue so he could spit blood at whoever was standing in the area. Eventually, he needed five people—one on each limb and someone to stand close enough with a towel to hold it up when Bryan sprayed blood."

The Alex Center handled each rage discretely, nonjudgmentally, and without emotionality. Slowly, the environment of care began to soften Bryan's reactivity. His runs scaled down to jogs, then walks. He would ask to do other activities instead of being upset with the current task. He excused himself to the quiet room if he started to feel shaky or uncertain. He even began to interact with other students in the program and developed new friendships. Where he once had three to four behavioral interventions a day, by year's end, Bryan had controlled his rages to only a handful a month.

. . .

Bill and Kathy's imbroglio with the school board wasn't in vain. They're now mentors in an organization that provides consultation to parents whose children have just received a serious medical diagnosis. If Bill and Kathy can save one family a few hours' worth of headaches and heartbreaks, then they feel they've made a difference. Nevertheless, they both readily admit that our educational system has to be reformed[4] if we're ever to provide adequate services to our brain-injured children.

Considering how many children she has seen move through the system, Sharon Grandinette has a privileged perspective. I asked her how she felt about Bryan's future.

"Bryan can legally stay in public school until the age of twenty-two," she said, "and it may take him that long. After that, he could enter into a vocational program, or he could learn a trade like welding or car painting—he has a natural gift for motors and engines. I think Bryan will be able to get a job and hold a job, provided the employer understands his learning curve. Once Bryan learns it, he's got it."

The outlook for Bryan seemed so much more positive than for so many other brain injury survivors, I had to ask Sharon if she felt Bryan had truly moved beyond the trappings of his damaged brain.

"His brain injury is not a hundred percent contained, it is only contained in the environments that we manage for him," she said. "The people around him have to know how to manage him. Let's say he goes to a math class, which he likes, and there is a substitute teacher and Bryan's aide is in the bathroom, and the teacher says start working on a math problem. You could pretty much figure out the interaction. It may revert back to the first day at the Alex Center, except now you have a six-foot-tall, two-hundred-pound kid with solid muscle, with twelve or thirteen other kids in the class who have learning disabilities and their own challenges. I can't predict what will happen, but I can tell you that psychosocial issues follow survivors for life."

In his ordered environment, Bryan Larson enjoys freedoms that would have been beyond his reach a decade earlier. He's in middle school now and gearing up for high school; he is reading and writing. He spends half his day in a brain injury program, and another portion of the day in public school. When Bryan first started at the Alex Center, his goals involved calming down and making it an entire day without exploding into a fury. Now his ambitions involve cars, hanging out with friends, and, yes, girls. He recently made the track team, no surprise there. It's been five years since he's had a seizure, and it's been at least two years since he's had a rage attack. Bryan Larson's other health impairment, his brain tumor, isn't likely to resurface. Although he's living in a managed world, Bryan hasn't felt this free in years.

I ask the Larsons how Bryan's brain injury changed their lives, and a quiet comes over the table.

"We lost four or five years of normal family time," Bill says, tears filling his eyes. "But I think it made us stronger."

FUGUE OF THE PONY SOLDIER

I'm lying bare-chested on the dirt floor of the *inipi*, with grass and pebbles stuck to my cheeks and forearms. Sweat is running from my head and back onto the ground in a hundred little rivers. I cup my hands around my mouth and take heaving drags of cool air from Mother Earth, who, I've been told, will save me in my hour of need. It's about a hundred and sixty degrees inside the sweat lodge—this is my hour of need. My head feels like a swirling cauldron, like I'm boiling my brain in its own juices, and my heart is pounding as if I've been running miles, although I haven't moved much for two hours. The Cherokee call this a humbling sweat. On my side, gasping, I am humbled.

Pony Soldier suggested the sweat lodge, and he wouldn't tell me much about it.

"What goes on in the *inipi*, stays in the *inipi*," he cautioned me.

The preparation for a sweat begins long before the first stone is heaped on the fire. First, a medicine man conducts a purification ceremony over the area where the *inipi* will stand. He dances, sacrifices herbs and tobacco, and chants his prayers. The hallowing ceremony requires several hours before the land is declared sacred ground and ready for the construction of the *inipi*. Twelve long, pliable willow limbs are pitched deep into the ground, forming a large circle of posts. More tobacco is sacrificed—this time in gratitude of the tree that furnished the limbs. The limbs are then bowed and fastened together at the top, with more limbs—the *inipi*'s ribs—secured along the sides. Covered with tarps and blankets, an *inipi* looks like an overturned basket with a flap that faces east or west.

"The west is where the storms come from," a Lakota brother tells me as we sit outside preparing to enter. "We face our *inipi* west to open ourselves to their power." His gleaming black eyes linger on me. He doesn't have to tell me that the west also represents suffering and death; I will feel it soon enough.

The grandfathers—numerous cantaloupe-size rocks from a nearby quarry—sit under a blazing bonfire. They're just turning from gray to white when the sweat lodge leader appears wearing a necklace of wolf claws. He's an old Dakota holy man with a healthy belly and silver hair pulled into a ponytail. He tells the five of us that the grandfathers will show us when they are ready to enter the *inipi*, and that we should be patient, as they have been waiting for us before the rivers formed.

The holy man kneels near the altar he arranged by the opening of the *inipi*. He asked me not to divulge specifics about the relics on his altar, but I can say that most traditional altars consist of either buffalo skulls or an eagle's feather. This holy man's altar upstages tradition: it is primal, mesmerizing, gruesome. During his prayers by the altar, he holds up a pinch of tobacco toward the west, then the north, then the south, and finally the east. He then beckons each of us over and fans smoking cedar around us, purifying us for entry into the sacred womb.

First the holy man enters, then the elders. As they crawl into the small opening, they each announce *mitakuye oyasin*, a phrase that translates as either "we are all related" or "all my relations." The words are not merely a declaration; they are regarded as the essential prayer, the mantra of the Plains Indians. In uttering "all my relations," one honors the sacred awareness present in all life and acknowledges it as the same awareness that flows within. On this ground, I was informed, we are all brothers. I file in behind a younger brother, lower myself on my hands and knees, and crawl into the dark.

The first time I saw Pony Soldier, he could do push-ups until you grew tired of counting them. He had the lowest resting heart rate I had ever measured on a healthy man, forty-two beats per minute, and the sinewy, lean muscles on his arms looked like they were made out of rigging cables. Poised arrow-straight, over six feet tall, and skin bronzed nearly red, Pony Soldier flashed a gold-toothed smile when he told me stories about his wayward childhood in the green reaches of Oklahoma's Indian reservations. The gold front teeth were unintentional but great conversation starters, he laughed, claiming they were the surprise result of a dental job he attempted to negotiate in Mexico.

Throughout his adolescence, Pony Soldier picked up a new musical instrument every year, and he ran track, wrestled, and would have gone to nationals with the swim team had they not demanded he shave his head. He also wrecked cars, dated the wilder girls, and refused to back down from a fight. Pony Soldier's untamed youth, however, left only traces of a prankster. Twelve men trusted him as their spiritual leader, three men trusted him as a foreman on a log cabin construction site, and his wife trusted him to keep the peace in a household prone to teenage outbursts. He was the kind of man who engendered respect from those who knew him.

Now, three years later, he's a broken version of the man I met. Where his once clear eyes flashed at the mention of his heritage, he now casts a sunken gaze in my direction. His dark mane is lost to a blasphemous, outgrown buzz cut, and his peculiar gold teeth replaced with dentures. Under the unkind fluorescent light, Pony Soldier's skin casts a sick greenish tone muted by his dark gray jumpsuit. He's only forty years old.

We're sitting in a visitor's room at the Lexington Assessment and Reception Center, a medium-security prison just south of Oklahoma City, where Pony Soldier is completing a shortened ten-year sentence for a crime he can't remember.

"This is where I was stabbed in the cafeteria," he says, raising his shirt to reveal a mean sliver of scar tissue below his right pec. He raises a pant leg, pulls down his collar, points to his hamstring, thumb-gestures behind a shoulder, holds out each forearm. In a matter of seconds, he's given me the stab-wound tour of his body, a total of eight scars he's received in the past three years. The last one punctured a lung and landed him in the hospital for two weeks.

"Right now, I'm on code level four, the best you can get here," he explains. The prison runs an awards-based behavioral modification plan, meant to dissuade violence. "One incident, and I start back at level one. It would take me a couple of years to get back to level four." Not that the days matter much to Pony Soldier. Ever since his brain injury, time has a tendency to jump around, to move backward and forward in jolts and skips. Calendars and clocks serve only as reminders of what he has lost, or what he's about to lose.

The Cherokee Nation's headquarters are based near the Illinois River in Tahlequah, Oklahoma, a long hour's drive southeast of Tulsa. Tahlequah was a final stop on the Trail of Tears, the horrific two-thousand-mile-long trek that in 1838 claimed the lives of up to four thousand Cherokee. Despite the tragedy they endured, the indomitable Cherokee spirit thrived alongside the Illinois River. Today, the Cherokee Nation is a federally recognized Indian tribe whose jurisdictional service extends over fourteen counties in northeastern Oklahoma. According to its own estimation, the Cherokee tribe now holds over two hundred thousand members.

Though many of their ancient sweat lodge rituals have been lost to time, Cherokee regard the sweat lodge as one of the sacraments of Native American culture. The white man, however, saw the sweat lodge as a societal threat, citing it in one circular as an "Indian Offense." In 1873, the federal government prohibited the ritual and systematically punished offenders. To settlers, the sweat lodge was an act of infamy not afforded the freedom of religious expression. Several tribes lost their sweat lodge rituals in the moratorium. Fortunately, the Lakota Indians preserved and propagated their sweat lodge liturgy, which now forms the basis for most intertribal sweats in North America.

In order to become a sweat lodge leader, a holy man must pass through his vision quest and gain fluency in the various prayers and songs germane to the sweat (a sweat can vary greatly in its focus and can incorporate healing rituals or peyotism). After a holy man serves some time as a firekeeper, a medicine man then bestows the right of leading a sweat lodge onto the holy man, who may then conduct his own sweats upon the request of his brothers. Pony Soldier regularly led sweats for twelve men, but some intertribal sweats can include forty or more participants. The numbers, however, are irrelevant. A sweat can occur in solitude, and often does. Because of the intense demands it places on the body, most leaders refuse to hold a sweat more than once or twice a month.

For the leader, the sweat starts three days before entering the *inipi*. Without any proclamation, he begins a fast and refrains from any ingestion of food or water. No brushing teeth, no bathing. The fasting period becomes a time of introspection and prayer, with thoughts often returning to the needs of his people. All other participants in the sweat are asked to fast one full day prior to entering the sacred womb.

Sweats generally last four rounds, but each round could take fifteen minutes or hours, depending on the movement of the Great Spirit. Participants may feel inclined to pray out loud or in silence, but an attitude of reverence and patience dominates the ceremony, no matter the intensity of the heat. The heat itself is regarded as a measure corresponding to the needs of the group. The leader who conducted my sweat informed me that during some sweats the grandfathers can stay red-hot the entire time, as a means of humbling the group, or they can cool quickly, showing their mercy toward a person in need.

People have died in sweat lodges. The heat can become so extreme in some sweats—particularly if there is a fifth "warrior" round—that metal jewelry will burn the flesh it touches and certain fabrics will melt onto skin. Steam rising from the grandfathers can blister foreheads and burn the throats and nasal passages of the unaccustomed sweatgoer. The risks inherent in a sweat lodge do not go unacknowledged. Several times prior to my entry, I was told that all I needed to do was to shout out the same phrase I was to use upon entering, "all my relations," and the leader would open the door to let me out. *Mitakuye oyasin* isn't just a prayer or blessing. It's also a cry for mercy.

"Sometimes I can't hear you, no matter how loud you are," the holy man warns me. "The spirits may cover my ears; they may not want you to leave."

In Cherokee, Pony Soldier's name is pronounced "Tatsa-dsa-si-go," a name he earned as a three-year-old who fearlessly played among the horses.

"It's a gift. The horses just accept him," his wife, Susie, told me.

She rattled off a list of tribes running in her blood, but Susie isn't a card-carrying member in any of them. Today, she lives in a small apartment in Pryor, Oklahoma, cashes a small disability check, and picks up odd jobs through her local church; her total income rarely exceeds seven hundred dollars a month. Their oldest son, Christian, has long since moved out, and their twelve-year-old daughter, Katelyn, is visiting a relative in North Carolina. It's a bittersweet respite for Susie. She misses her daughter, but the times are hard and her daughter's absence offers a summer's worth of financial relief.

I asked her what life was like before Pony Soldier's accident.

"Before the injury, the days are full of work," she said, her voice heavy with longing. Susie told me they were restoring a beautiful three-bedroom lakefront home that summer, with the hopes of completing it later that fall. "When we get some family time, we do some remodeling on the house or go to the river. And music. There's always music playing somewhere."

Family time, at that point, meant weekends. Pony Soldier had picked up a subcontracting job to build a lakefront log cabin on Grand Lake—two hours away. Instead of driving home, Pony Soldier wrung the labor from every ray of daylight and crashed at a nearby campsite. If he could work twelve-hour days, he thought, he could have the job finished early, under budget, and in time to pick up another contract before summer's end. Weekdays were workdays, plain and simple, but Saturday mornings were sweeter still.

If he wasn't leading a monthly sweat, Pony Soldier was up at five in the morning, preparing for his day with a routine cup of coffee. At six, he played fetch with Cherokee, and after he'd worn out the dog, he would wake up Katelyn.

"She's a daddy's girl if ever there was one," Susie said, pointing to a picture of Katelyn on the wall. She's the spitting image of the Pony Soldier I knew: eyes wide and bright, a sharp nose, dark skin. "She was just six when he got the injury."

Pony Soldier is down by the shore when it happens. There is a pile of logs meant to be support beams in the master bedroom, and Pony Soldier grabs the end of one log while his coworker, an eighteen-year-old apprentice, grabs the other end. Instead of allotting for a wide turn that would let Pony Soldier walk around the log pile, the kid heads straight for the house, forcing Pony Soldier to scurry over the stack. Midway over, the logs loosen under Pony Soldier's weight. He tries to keep his balance while the logs roll out from under him, but he loses his footing.

A split-second decision both saves Pony Soldier's life and changes it irrevocably. Falling toward his back, he tosses his end of the log up and thrusts his body out from under it. Pony Soldier slams onto another log, then attempts to duck the thrown log. The end of the log hammers down on Pony Soldier, clipping him just behind his right ear and thumping his head against the logs below.

"They said I couldn't get up or nothing," says Pony Soldier. "I was disoriented and dazed, and they pulled me off the logs and set me under a tree with a jacket underneath my head."

The contractor gazed into Pony Soldier's eyes and gave his diagnosis, based on pure anecdote. He told Pony Soldier that since his eyes weren't red, he must not have a concussion, so there was no need for medical attention. Red eyes or not, the actual symptoms of a concussion include headache, dizziness, disorientation, loss of memory, and vomiting. Shortly after the incident, Pony Soldier began vomiting and complaining of severe head pain. The contractor offered him a bottle of ibuprofen while he and the other workers packed up the tools. Pony Soldier munched the pills like candy.

Pony Soldier continued working for two more days, but his coworkers noticed a change the following morning. Anytime the radio came on, Pony Soldier rushed to turn it off. Normally gregarious and high-energy, he hardly said a word to anyone else on the job and ran toward the bushes every time he needed to vomit. The bottle of ibuprofen barely lasted him a day.

The accident happened on a Wednesday in May. By Friday, Pony Soldier couldn't stand the pain much longer. He knew he needed medical attention, despite what his coworkers told him.

"He called saying he was on his way to Hastings [Indian Hospital] but did not go into detail," Susie said. "I knew something was wrong because he'd do anything to avoid a doctor's visit. He told me he would be home soon, and that was the last I heard from him."

"All my relations," I say, and duck my head into the *inipi*'s opening.

The first thing I notice when I'm inside is how cool the dark feels. I take a place, roughly eight o'clock to the holy man's midnight seat, which is directly opposite the *inipi*'s opening. In the center of the *inipi* there's a fire pit, an almost perfect circle in the dirt, about six inches deep. The holy man grunts a command in Lakota to the firekeeper, who disappears and then returns with a grandfather in between the iron pinchers. He does this four times, until four lone rocks sit in the pit, forming a diamond shape. I can feel their warmth radiating toward me, and I stare at them, wondering if they can smell my fear.

The four rocks are the four directions, and each one receives a whispered blessing and a sprinkle of tobacco. With another grunt, the rest of the grandfathers are ushered in, one by one. Each rock is about the size of a cantaloupe and looks bone white, and each rock makes its presence known with the heat it brings. Another grunt closes the *inipi*, and light is completely blocked from entering. It takes a moment for my eyes to adjust, but the first thing I see is the orange glow of the grandfathers. A bead of sweat trickles down my temple.

I tried to prepare for it. I lasted ten minutes in a steam room at the Hotel Renaissance in downtown Des Moines, and I made it twenty-five minutes in the dry sauna at the Sheraton of Anchorage, where the temperature was a comparatively mild one hundred and thirty degrees. My friends and I once drove through Death Valley in an open jeep, but that was ten years ago, and we stopped every half hour to soak our beach towels in water and wrap them around our heads and shoulders. Death Valley loses its sting when paired against a sweat. The temperature in an *inipi* can simmer near two hundred degrees, a temperature more suitable for baking a ham than sweating.

Only the outlines of my five brothers are visible against the fire pit's glow, and for a few quiet minutes, only the grandfathers make a sound as occasional tufts of tobacco burst into small flames, giving the air a faintly sweet smell. From a corner, the drum starts beating and the holy man begins a low, mournful song. The glow, the drum, and the deep-voiced chanting invoke the most primitive atmosphere I've ever witnessed.

A loud crack and sizzle punctures the darkness and suddenly a wave of sage-scented steam engulfs my face. The heat rushes down my body and I hardly have time to take a breath before another hornful of water gets dumped on the rocks. Already the air is starting to feel thin and insufficient. The leader continues dumping more and more water onto the grandfathers, and each burst of steam ratchets up the heat more and more. I'm breathing deeper with almost each inhalation. As the steam moves in and out of my mouth, the skin around my lips and nose feels like I've rubbed it with chili peppers. The first round isn't even half over, and all I can think of is "all my relations."

Just then, the steaming stops, and the leader sets the horn aside. He speaks to the Lakota brother in his own tongue, then the brother responds with his prayers in English, a gift to the rest of us. During the first round, prayers are intentioned toward personal needs and should not concern the welfare of others.

"Strengthen yourself before you pray for yourself," Pony Soldier told me. "You are there for strength, because our church, the *inipi*, takes care of all aspects of man—the emotional, the spiritual, the physical."

I listen to each of my brothers bare their hearts, addressing in turn the Creator, God, the Father, the Mother, the Great Spirit. Their prayers are beautiful and earnest and it doesn't occur to me until it's my turn that I have forgotten the heat. As sweat begins to pour down my neck and arms, I speak my intentions, entrusting them to the same swirling steam as my brothers have done, just as thousands of their brothers have done throughout the ages. Our prayers, the prayers of all my relations, become one in the timeless heart of the *inipi*.

After a period of silence, the leader sings another prayer, this time at just a whisper, and only for a few minutes. When he calls out, the firekeeper folds the western opening of the *inipi* back and cool air and blue light flood the floor of the room. The initial round is over. Each of us has weathered the passage, and we all recline onto our elbows to savor each fresh breath. The holy man passes around a horn full of herbed water. I drink part of it, pour part of it over my shoulders, my head, and my back, and save the last drops for the grandfathers. In a few moments, the opening will once again be sealed, and we'll begin another, more intense round, a round not everyone endures.

Six days following his injury, Pony Soldier's memory started again, only it picked up more than a year in the past. He was unable to sense any order, as if the chronology of his life was suddenly scrambled. He looked around and figured he was trucking for a living. He was at a road stop in Denton, Texas, and when he went outside to get back in his new eighteen-wheeler, it was missing. He scoured the parking lot for it in vain. Pony Soldier told himself that he had to get home, except he decided home must be New Mexico, where he had lived before moving to Oklahoma. For some reason, he couldn't remember a single phone number, and with his belongings in the truck, he reckoned he was stranded. He walked out of the parking lot, not recognizing his actual vehicle parked in front of the store. The nearest place to Denton was his sister's house in Moore, Oklahoma, so he began his trek north on foot, right up I-35.

Pony Soldier crossed the Oklahoma state line before an officer pulled up behind him and asked him his business.

"I'm just trying to get home to my family," he explained. The officer took a long look at his dirty jeans and tangled hair and figured that he was just some drunk, homeless Indian who needed to dry out in the tank. He told Pony Soldier to get in his car, that he would help get him home, and that they needed to stop by the jail first.

While Pony Soldier started dialing for directory assistance in New Mexico, the officer took his wallet and ran a check on his identification. Pony Soldier turned up on a missing person's report, and the officer called Susie, who nearly burst into tears on the other end of the line. She begged the officer to keep him in jail until she arrived, and explained that he'd hit his head at work. Three hours after leaving her daughter with a relative, Susie arrived with Christian at the Love County Police Department. When she finally saw her husband, Susie could already tell the difference.

"He was quiet and confused," Susie tells me. "I thought he was going to cry, but then he asked me if he had ever hurt me, and I told him he hadn't. Then he wanted to know why our phone was disconnected, and I said it wasn't. I finally got around to asking him what happened to our Blazer."

"What Blazer?" he asked.

More than a year was lost to Pony Soldier, and he laughed when Susie tried to convince him otherwise. It must be a joke, he thought. He couldn't believe that they were living in Oklahoma, that he was no longer a truck driver, and that he had returned to a career in carpentry. But he also couldn't explain how he made it to Denton, or what happened to his truck, or how he turned up on a missing person's report. Most of all, he couldn't explain the disturbed look on his son's face. He agreed to get checked out at the nearest Indian hospital, Carl Albert in Ada, Oklahoma.

"They flat told me he had a head injury," recalls Susie.

What they didn't tell her was that Pony Soldier was experiencing a bizarre neuropsychological phenomenon that can shift diagnosis depending on the setting. A general practice doctor is likely to label a similar case as simple amnesia secondary to concussion. A psychiatrist might arrive at dissociative fugue disorder, an episodic event wherein a person disconnects from their current sense of identity, takes flight from their home, then wakes up hours or days later without any memory of the journey. Dissociative fugue is usually attributed to a psychologically traumatic event, such as extreme abuse. Treatment might focus on Pony Soldier's family history in hopes of finding a psychogenic source. An inquisitive neurologist or neuropsychologist, however, might have observed that Pony Soldier's experience did indeed involve an amnesic state, a dissociative state, and a fugue state, but the picture wouldn't be complete without the preflight symptoms of headache, left-side weakness, and nausea—symptoms that smelled like the makings of prolonged absence seizures. Traumatic brain injuries are notorious for their ability to slip between our current diagnostic understandings.

At Carl Albert, Pony Soldier was X-rayed and examined by a doctor in the emergency room. Not a psychiatrist, not a neuropsychologist, not a neurologist. The resident explained to Susie that her husband had experienced a traumatic brain injury and needed medical attention, and that he could stay there in Ada, but it was so far away. Wouldn't it be better, the doctor suggested, if you just took your husband closer to home? Susie accepted the sample packets of medications, then took her husband by one hand, her son by the other, and together they drove back home that same night.

I ask Pony Soldier what the drive was like for him.

"I guess you could say I was half in the past," he tells me. "I was different in my thoughts and memories. To be honest with you, my life hasn't really all come back."

Pony Soldier's first flight caused him to second-guess his own memory, to distrust knowledge he once regarded as inexorable and enduring. If the calendar said June, he checked the year to be sure. He stumbled on skills he didn't know he had, like replacing a carburetor. He couldn't afford a cell phone, so if he had to go anywhere without Susie, he made sure she knew his every step. It was Pony Soldier's second flight, however, that pushed him past the boundaries of helplessness and into the domain of fear and suicide.

The night before the Battle of Little Big Horn, a twelve-year-old Lakota boy painted his body yellow, smeared black stripes over each eye, and fastened his hair up with eagle feathers. In the ceremony that followed, a holy man called upon the spirit of the bear to deliver a healing power.

"Suddenly, I felt myself lifted clear off the ground," Black Elk recalled. "And while I was that way, I knew more things than I could tell, and I felt something terrible was going to happen in a short time. I was frightened."

The following day, the young Black Elk shot an arrow through the forehead of a cavalryman and scalped two other soldiers. In the ensuing battle, called the Battle of Greasy Grass by the Lakotas, a combined group of Cheyenne and Lakota Indian warriors annihilated General Custer and 267 soldiers of the Seventh Cavalry of the United States in less than three hours. The estimate by the National Park Service places the Lakota losses at around forty. For Black Elk and the northern plains Indians, the battle initiated a bloody struggle for survival across the Great Plains. Though he earned a reputation as a fearless warrior, the little fighter at Greasy Grass grew into a *wichasa wakan*, a holy man revered by all those who knew him. Some fifty years later, Black Elk gave his account of "The Rubbing Out of Long Hair (Custer)," along with many other life stories, to the poet John Neihardt. Today, Neihardt's conversations with Black Elk, along with Joseph Epes Brown's *The Sacred Pipe: Black Elk's Account of the Seven Rites of the Oglala Sioux* stand as two of the most important records of the religious rites and prayers of Native Americans.

Of the second round of sweating in the *inipi*, Black Elk explained that the power of the northern eagle is invoked during the opening prayers. Before I have a chance to recuperate from the first round, the door of the *inipi* is sealed once more, and the darkness is more ominous than before, just as the heat in the lodge begins with greater intensity. In the dark, the grandfathers take on an angry, invigorated glow. The holy man's wail sends a chill up my spine, and once again, the phrase "all my relations" sits perched on my lips. The round starts without one of the elders, who excused himself due to heart problems.

The air fills with thick, dry cedar smoke and I start coughing and hacking in the dark. I don't feel so bad when I hear another brother cough as well. The hot smoke is merciless and burns my eyes and chaps my throat and I keep coughing until I get a face full of steam. The second burst makes me gasp out loud. If my brothers could see the agonized look on my face, they'd probably expect me to cry. The perspiration running down my face feels like tears. Sweating this hard is a form of crying.

"All my relations!" shouts a younger brother, nearly taking the words out of my mouth, and the holy man yells back a command. The door of the *inipi* cracks open and I see the glistening blur of one of the brothers my age duck outside. Just as quickly, it's dark again. With two men out, a third of the attendees are gone now and the sweat isn't half over. The steam continues with a vengeance and my breathing gets deeper and deeper until I hear the wailing once again fade into a whisper.

"The second round of prayers is for all those with authority over us," Pony Soldier told me. "Whether it's your boss or your homeowner's association. You can pray for everyone under the sun who is over you, plus soldiers and protectors, wardens and judges. You won't find a lot of brothers who will admit to praying for their wife, though a bunch of them do."

My eyes take notice when I hear one brother praying for his mother, but then it makes sense. The Cherokee are a matriarchal tribe, where women can share chiefdom on par with men. Each of the remaining brothers says his prayers out loud, and I'm embarrassed by their sense of place within their community. They acknowledge the people in their lives you wouldn't consider authorities: best friends, complete strangers, even their enemies and oppressors. When it's my turn, I'm struck silent, and so I go with the stillness and offer my prayers wordlessly. The room is quiet save for the sound of heavy breathing, which is happening in unison, as though the *inipi* itself has one breath.

By the time Pony Soldier underwent examinations at both Indian hospitals, he had accumulated a gym bag full of prescription drugs. Although various doctors listened to his symptoms, none of them offered more than cursory pain management for his headaches and antinausea pills for his persistent vomiting. Hastings, the hospital nearest to home, told him that he needed a referral from a primary care physician before he could get in-depth treatment. Pony Soldier bounced from clinic to clinic in hopes of finding an understanding ear, but the one Indian health doctor who could authorize his hospitalization had a month-long waiting list (healthcare information and access are major areas of concern for Native Americans with brain injuries, according to the Indigenous People's Brain Injury Association).

Pony Soldier had attempted to go back to work on the cabin but lasted only a half-day before the retching overtook him. He was already starting to notice that heavy physical exertion seemed to bring on the attacks. While on the lookout for less demanding work, Pony Soldier tried to return to remodeling his house. He figured if he could go at his own pace, he could hold the headaches off longer.

"Before, I would watch him dance and sing while he worked on the house," remembers Susie. "He could throw up a piece of Sheetrock and not even think about what he was doing. One day [after the accident] I was watching him work on the house, and he was just standing in the middle of the room with his tools in his hand, staring at a panel of Sheetrock. The hammer just slipped out of his left hand onto the floor. He had tears in his eyes and he told me that he had forgotten what he was doing. That time, it hit all of us hard."

When Pony Soldier finally did get in to see his primary care physician, the nausea had finally slowed from several vomiting episodes a day to only a few a week. The headaches, however, persisted. She wrote him a prescription for narcotic painkillers and told him to check back with her if the headaches didn't go away. Pony Soldier figured that if his primary care physician didn't see a need for him to get further testing or hospitalization, he would just have to come to terms with the pain. He despised how painkillers made him feel and hoped the headaches would eventually go away.

By the end of July, Pony Soldier's mailbox presented a new bill nearly every day. The stress fed his headaches, and the lack of work fed his stress.

"He was pacing like a caged animal at that point," says Susie. "There was no more music in the house, and every once in a while, he would just go white and that's when you knew he was going to start throwing up. It got to be a regular thing."

In their quiet ways, the locals in the small community offered Pony Soldier small jobs repairing fences and fixing loose doorjambs. Although he wasn't quite sure it was a good idea, Pony Soldier accepted an offer to mow the lawn of an elderly churchgoing couple. He loaded up his Blazer with the mower and a gas can, drove to the couple's house, and unloaded the mower. The next thing Pony Soldier can recall is waking up in the middle of a campground.

"I was lying down beside my truck," says Pony Soldier. "I was clean and I didn't look like I'd been in the mud or nothing."

He took a look around the campsite and saw a smoldering fire and several days' worth of food wrappers and empty bottles. If his earlier blackout seemed like a jump into the past, this one sent time splintering into fragments. The confusion was unbearable. Where had he been? Why was he camping? Had something happened to his family? Did he still have a family? Each question only spurred more questions. Eventually Pony Soldier reached into his pockets, felt a set of keys, climbed into the nearby truck, and drove until he found the name of the town: Eufala. The knowledge caused a landslide of connections. Eufala was ninety miles from home; home was Oklahoma. He no longer lived in New Mexico or California. Try as he could, though, Pony Soldier could not figure out why his head was pounding.

"I thought I was going mad," Pony Soldier says. "And I got depressed. I felt like I was costing my family too much."

The pain and the chaos ravaged his mind. Whoever he was, he felt wholly useless and in agony. When he drove over the train tracks, an impulse overtook him. He pulled his truck onto the shoulder of the road, walked a few paces down the track, lay down across the ties with his neck over one of the rails, and waited. He clenched his teeth and cried, half from the searing pain in his head, half from the madness. This is no way to live, he told himself. There is no honor in this.

The train never came. Pony Soldier lifted himself off the tracks and climbed into his truck.

"He called from a homeless shelter," Susie told me. "He begged me to come get him before he had another headache."

Susie picked her husband up and once again detailed to him how he had been working on a job site and injured his brain. He was prone to blackouts now, she told him, and this was the second time he'd been on a missing person's list in three months. During this episode, a friend of theirs had spotted Pony Soldier leaving a convenience store in Eufala and called Susie. When Susie arrived there with a picture of Pony Soldier, the clerk told her that she recognized the man because he came in for coffee every morning that week. He was part of a road construction crew, she said. The day before Pony Soldier called, Susie had scoured the county without luck.

When they arrived home, they shared the same mindset: something had to be done. They also knew it had to be done without any money. From her conversations with social workers, Susie learned about the federal law that requires emergency rooms to treat patients who are suicidal, whether they have insurance or not. She drove her husband straight to Hillcrest Medical Center in Tulsa and had him committed to the psychiatric crisis unit, a locked-down ward. Both she and Pony Soldier knew it wasn't exactly the right place for him, but at least he would be under observation there.

Hillcrest is one of Tulsa's largest medical complexes, and while there, Pony Soldier was finally able to get an initial MRI done. The results came back indicating a 1.5- by 1.3-centimeter mass buried near the medial anterior edge of the right temporal lobe. Although the mass was decidedly benign, its exact composition remained unclear. The MRI technician recommended further scans in order to find out what seemed to be encroaching on Pony Soldier's temporal lobe.

In 1998 a group of Italian researchers documented several episodes of a dissociative disorder in a thirty-two-year-old man following his traumatic brain injury. They reported that forty-five days after his accident, the patient experienced a "progressive reduction of awareness . . . coupled with a trance state followed by amnesia of the event." During these episodes, the patient was unable to integrate various aspects of his memory, consciousness, and identity—a suitable description of Pony Soldier's own experiences. The researchers concluded that those "who deal with TBI should be alerted to this possibility [of dissociative disorder] which may be less rare than previously thought."

I had a chance to observe Pony Soldier during his two-week stay at Hillcrest. Patients are kept on an acute psychiatric unit because they are at the most extreme point of distress in their lives. A handful of them are going through psychotic breaks and are actively hallucinating. Many of them still have charcoal stains around their lips from having their stomachs pumped after an overdose attempt. They walk in stupors or huddle in corners, and they look tired and despondent. It's a hellish environment, one in which Pony Soldier stood out like a sore thumb.

During the most confusing months of his life, Pony Soldier stood upright and talked with clarity and precision; he carried himself with uncommon dignity. Within a few days, his headaches were medically controlled, he slept well, and confusion no longer held him in its grip. He seemed determined to reclaim his sense of order. He called his blackouts "time lapses" and expressed concern about the ramifications of his condition. How, he wondered, could he be expected to be a good father and husband if he could disappear at any moment? Was he a different person while dissociated? He had heard Jekyll/Hyde stories about people who switch personalities, and he worried that his blackouts might allow his inner demons their license. In his words, the condition had "taken him far from his *inipi* circle," meaning that his injury had pulled him away from his spiritual practice, and he hoped that soon he would be able to return. Without expecting it, Pony Soldier was about to receive an answer to one of his questions.

A few days before his discharge, Susie appeared during a visit and leveled Pony Soldier with shocking news: criminal charges. A woman living near the Eufala area had filed a police report claiming that two men arrived at her house asking to do laundry. She allowed them in, but while there, the report stated, the two men stole her computer along with several pieces of heirloom jewelry. The license plate on the vehicle matched that of Pony Soldier's truck, and the description of one of the men matched Pony Soldier. When he was discharged from Hillcrest, Pony Soldier would have to answer to felony charges for second-degree burglary.

It's the beginning of the third round in the *inipi*, and I feel weak and lightheaded. My blood feels as thin as warm water rushing through my veins and I notice myself moving around a lot. Comfort evaporates at a hundred and fifty degrees. The steam is a succession of stinging, wet slaps. My jaw is slack and I'm breathing from my heels. I'm convinced there's no real oxygen left inside the *inipi*, just smoke and steam.

The chanting and the drums begin to sound far away, on a different plane. Against my wishes, my elbows collapse onto the ground and my head follows. Droplets roll off my cheeks and my chin drips like a coffeemaker. Six inches above the ground, the air is slightly cooler, just enough to help me get past the fainting feeling. I take a few deep drags and look around to see that I'm not the only one having a personal moment with Mother Earth. I get the distinct impression that I'm paying for something, that through sweating my body is atoning for actions my mind can't reconcile. It's painful, and yet I'm not alone in the pain.

The prayers in round three are petitionary pleas on behalf of family members, but as I hear the words coming out of one brother's mouth, I notice he's praying but not really asking for anything. Growing up in the buckle of the Bible Belt, I heard preachers ask God to remove tumors and vanquish psychological guilt, and I heard pastors beg Jesus for new church gymnasiums and fleet vehicles. I'm accustomed to prayers born of ego, where the will of God and the will of the supplicant are deemed interchangeable. My brother in the *inipi* prays for acceptance, as though he trusts that the will of God will be done regardless of what he asks. He prays that a sick relative comes to terms with her sickness, even if it means death. He prays that a child opens his eyes to the poverty of his surroundings. This is Indian atonement. He prays not to manipulate the world to his preferences, but to embrace it in its imperfection.

There is barely time for a few whimpers once the prayers subside. The opening of the *inipi* is cast wide for only a moment, as though it's a mere gesture, a reminder that you can get out now, but if you go, you go alone. I stare longingly at the light beyond the opening and don't take my eyes off the door until it's closed. The holy man asks if everyone is all right, and by everyone, I know he means me. I nod my head and prop myself up into a sitting position, and the door closes for the last time.

When Susie found out that Pony Soldier was going to enter a plea of guilty, she was furious with him. Although none of the stolen items were ever found in his possession and his accomplice was never located, Pony Soldier felt he had no rebuttal, and after speaking with a public defender he was convinced he would be found guilty. The truth about his past embarrassed him: he could not say he was incapable of theft. In his youth, he had stolen his father's car and his father pressed charges to teach him a lesson. He already had a felony conviction for larceny, and it wouldn't take much for a jury to make the leap to burglary—especially if he had no memory of the crime. There wasn't any money for a proper defense, and Pony Soldier reckoned that the prison system might even be forced to treat his injury. He even felt relief knowing his absence would lessen the financial burden on Susie and the children.

The ensuing legal matters left no money or time for follow-up medical care. The recommended tests were abandoned and the packets of sample medications soon ran out. That November Pony Soldier was sentenced to ten years for second-degree burglary, and through a series of petitions and recommendations, found his sentence commuted to three years with probation. Soon after he was locked up, he began to complain of headaches, but the prison doctor denied him any medications, suggesting the pain might be relieved by a little dental work. In a fit of desperation, Pony Soldier agreed to the total removal of his teeth. The extractions had no effect on his headaches. At least now he had a set of white dentures to replace the row of gold teeth he'd worn for the past ten years.

The Lexington Assessment and Reception Center houses enough Native Americans to warrant a sweat lodge on its grounds. On his first attempt to sit through a prison sweat, Pony Soldier discovered that he no longer knew the words to any of the ceremonial prayers or songs. He sat quietly throughout the sweat, embarrassed by his inability to fully participate as he once could. He attended the sweat lodges only a handful of times after that, noting that they always initiated a terrible headache. The stresses of prison life, along with the hard labor it often demanded, kept his headaches active. He rarely went a week without one.

The meticulous regularity of the prison schedule caused his sense of time to crumble. During one dissociative state, he awoke several weeks later, wondering what had happened to the entire month of August. Although he made it a strict personal policy never to discuss his brain injury with any fellow prisoners for fear of appearing vulnerable, Pony Soldier asked his cellmate if he had been acting differently or strangely in the past few weeks.

"Now that you mention it," his cellmate told him, "you've hardly said a word and you've been staying in bed a lot. I just figured you were done talking for a while."

When I meet with Pony Soldier, he tells me he is scheduled to be released soon, and that he is planning on joining his wife and daughter at their apartment in Pryor. I ask him, in all earnestness, if he thinks he can ever have a normal life again.

"God, I hope so," he says. "Not having a normal life is my number one fear. First thing I want to ask my parole officer is if I can get some sort of tracking device on me that can show where I am so my wife can find me right away. I don't want to end up back in prison. My wife said something to me that I was scheduled for surgery tentatively when I get out, but what do I do to protect myself until then? Do I chain myself to my house and get a job on the computer, just so I can have an alibi? This is a real concern to me. I don't have a memory of my brain injury, and I don't have a memory of my crime. How does someone deal with this?"

"I don't know," I tell him. "The book hasn't been written on this kind of thing. You're going to have to tell the rest of us how you deal with it."

"Anything goes in the fourth round," Pony Soldier told me. "Your heart will tell you what to pray about."

My heart tells me to pray that I make it out of the *inipi* without heat stroke. The holy man goes into the fourth round with renewed gusto, and his opening song sounds like a battle cry that intensifies as the drumbeat hastens. I hear the other brothers join in the song, their voices sounding strong and defiant, as though they're preparing for war. The rocks take on a bright glow, as if they're feeding off the voices. I'm clenching my teeth and my fists in preparation.

"O You who control the path where the sun comes up, look upon us with Your red and blue days," the prayers demand. "O You who have knowledge, give some of it to us, that our hearts may be enlightened, that we may know all that is sacred."

The bursts of steam come on strong and fast. My brothers are all ululating and if I knew how to ululate without sounding like a shrieking little girl, I would join them. Each blast feels like my face is being dunked in boiling water, and there's barely time to gasp for air before another wave hits me. I can't distinguish between my insides and outsides; I'm not sure where the sweat stops and the steam starts. My eyes are clenched shut and I keep thinking that the steam has to end soon, but it doesn't. The holy man has to go through all the water, hornful by hornful until it has all been sacrificed to the grandfathers.

Somewhere in the steam, I give up. I give up thinking that I can make it. I give up caring what my brothers think. I give up expecting anything from this *inipi*. I give up on my house and the hedges I need to trim in the morning. I give up on this fucking job with all its fucked-up brains that must be saved and can't be. I give up on healthcare reform, I give up on advocacy, I give up on hospitals and nursing homes and institutions and prisons. I give up on my boss, I give up on my wife and my two daughters, I give up on my brother and sisters and mother. I give up on everyone I love and I give up on everyone I hate. I give up all my relations. I give up until there's nothing left to give up.

The door of the *inipi* opens to the gentlest light. My brothers and I file out wordlessly, our bare torsos shimmering in the red Oklahoma dusk. We sit shoulder to shoulder on a log while one of the elders stokes up the peace pipe. I take a few tugs when it is passed. One by one, we make our way down to the riverbed and enter the cool waters. I walk out to the middle of the river, lower myself to the neck, and face the current so that it lifts my arms and fingertips near the surface. The river's flow feels strong and healthy, and before I let it take me under I think of all my relations, all my relations, all my relations.

Susie is finally getting to a place where she seems comfortable with the uncertainty, and I blow her mood by asking her what kind of future she plans to have with her husband. Her eyes are angry and sad when she looks at me.

"I feel like screaming because of the chaos the injury has brought into our lives," she tells me. "We are told we can get help, and we don't. Since this happened, the majority of our discussions involve trying to piece things together. We can't relate like before, we can't have the same kind of conversations. All I want is my husband back."

I can't offer much in the way of consolation to her. She asks me if Pony Soldier qualifies for treatment at my hospital, and I tell her no. I explain that he doesn't fit our profile, that he is too high-functioning, too in control for our program. He doesn't fit our criteria, and he doesn't fit any placement that I know of. Our healthcare system doesn't serve him, so the best he can hope for is outpatient checkups at the local Indian hospital. With some luck he might be able to get a referral to a neurologist, who would probably put him on medications that he couldn't afford. Susie looks at me as if she wants to hit me, and I can't blame her. It might make us both feel better if she did.

IN ALL EARNESTNESS

Joshū earnestly asked Nansen, "What is the Way?"
Nansen answered, "The ordinary mind is the Way."

—FROM CASE 19, *THE GATELESS GATE*

So much snow, so much snow. Melissa rode in the backseat, watching the night whirl white with its first snowfall of the season. As the car she rode in passed slower vehicles, the wheels pulled and crunched and slid against icy patches. The driver, a doctor named Darrell, wasn't accustomed to Thunder Bay's blizzards—he drove as if he were back home, hundreds of miles south. He, Melissa, and two other passengers had carpooled to a housewarming party together, and Melissa's condo downtown was the first stop on the return home. She directed Darrell to take the next exit because it was a safer route.

Up ahead, on the hill, a pair of headlights flashed on and off again repeatedly. As the lights drew closer, Melissa could see the pair of lights sweep away to the left before reappearing moments later. The car ahead was lost in pirouettes, like a child's top, awkwardly spinning out of control and careening toward their vehicle. Darrell didn't have enough time to slow, but he managed to yell out to his passengers.

"Brace yourselves!" he cried.

Melissa lunged against her restraint and her hands flew instinctively around her face. Glass sprayed over coats and scarves. The impact pinned her, unconscious, in the backseat. The unbelted woman in the front seat crumpled under the dashboard. When Melissa awoke a few moments later, it was as though the whole of reality were occurring inside a snow globe. Everything was quiet and slow, so slow it might be death, she thought. She sat in the backseat and told herself that if she were dead, she wouldn't be looking at the wreck through her own eyes, and yet she didn't feel quite alive. Minutes later, though, the sense of life returned, along with pockets of sound and time and touch.

More than ten years later, Melissa Felteau's world is still returning to her, bit by bit. She no longer concerns herself with its fragments. Melissa doesn't long for the world she left that day; she is pleased, maybe even in love, with the world before her. For now, that world happens to be Ottawa, Canada, where she's conducting research aimed at helping other brain injury survivors cope with their impairments.

When Melissa greets me with a hug at the airport, I see nothing of the woman she was supposed to have been before the wreck. I'm usually uncomfortable around type-A corporate executives—the competitiveness and the moneyed optimism is all too much. Melissa radiates a warmth independent of station and geography. She's overly bundled in a camel hair overcoat. With her blonde hair and rosy cheeks, Melissa looks like the friend you want to call after good news or a hard day. Right away, you get the impression she is on your side.

We have a break before her research group meets, so we walk around a bit downtown and find a quiet restaurant for lunch. The brain injury is still there, but only when you're looking for it. Melissa talks in loops, circumstantially, dancing near the topic. She tells me that the driver in the other car began wandering the scene barefooted and dazed, and she likes this place because the food is great but the waiters are delicious. An off-duty emergency responder pulled over and offered a hand—remind me to tell you a story about him, Percy, the responder, I saw him years later, she says. Percy said that there were wrecks all over town and to expect a long delay for an ambulance. Thunder Bay is so rural, you know. The woman under the dashboard, Darrell's poor, poor mother, was making a wretched noise each time she exhaled, causing everyone to wonder if the following breath would be her last, so, no, I did not focus on myself. Darrell was hanging halfway out of the car. Percy made a quick assessment of the scene, and when he noticed me sitting calmly he put in a call requesting priority service. I was in the worst trouble: I didn't complain of any pain.

The more Melissa talks, the tighter the loop gets, as if her brain is working to reel itself in. After a few minutes, she's locked on topic. Knowing the extent of her injury, it's an impressive feat.

"I kept going within and within," she tells me. "I felt less able to breathe, like I was fading away."

First a fire truck arrives. A fireman hops off the truck and begins diverting the traffic for the ambulance's arrival. Another fireman looks into Melissa's eyes and straps an oxygen mask to her face. When the ambulance finally appears, forty-five patchy minutes have passed. Melissa watches calmly as they strap her to a spineboard and immobilize her neck. In the ambulance, they grimace as her vital signs weaken. A tech tells her to hold on for five more minutes, and she does.

The buzz and brightness of the emergency room jolts Melissa awake. In a matter of seconds, she is stripped nude. One breast looks as though it has been ripped off, and X-rays reveal four broken ribs. Her abdomen is swollen from internal bleeding, a ruptured spleen. Get her stabilized, go, go, go. Melissa is injected, bandaged, and whisked away, and anything that might have happened to her brain is a mere afterthought. She had survived, she would be fine, she tells herself.

Melissa spends Saturday spraying vomit into her bedpan and vehemently denying anything is wrong with her. She laughs and cries and rages within moments of one another and complains that she cannot see the ceiling. Her parents, exhausted from the overnight vigil, wonder how much morphine the doctors have given Melissa to cause her to act this strangely. Eventually, a resident on his rounds takes notice and schedules lab work and a consultation with a neurologist for Sunday, just to be sure.

A CAT scan returns inconclusive, so the neurologist delivers a vague but distressing diagnosis: postconcussive syndrome (PCS). Some of the most tragic cases I have seen resulted from a PCS. Because a person can move about and consciously respond, they're presumed to be well when they're far from it. The adverse symptoms snowball over a course of days. In the impact, Melissa's position in the vehicle made her prone to a concentration of forces that shook, turned, and rattled her brain, causing widespread disruptions on microscopic levels. Detailed scans a year later will reveal that Melissa's brain was riddled with focal defects, pockets of arbitrarily traumatized tissue encapsulated within healthy brain. Sheering injuries in her occipital lobe disrupted key functioning in her visual cortex; diffuse axonal injuries severed untold neuronal connections within the right frontal lobe, compromising her self-regulation. Cellular injuries to her parietal lobe interrupted the normal processing of sensory input. The microscopic defects elude the CAT scan, however, and consequently the neurologist. He writes down his diagnosis and offers no follow-up care, not even the hint that she should keep an eye out for any life changes.

Prior to the accident, Melissa Felteau was a paragon of modern womanhood, the embodiment of every contemporary virtue.

"I was a real catch," she says, giggling a little at her own cheekiness. "A healthy northern girl."

As a master swimmer, she outpaced men twice her size; on cross-country skis, she sailed past teenagers; and in whitewater kayaking, nobody could rival her quick rolls. Yet her athletic skills were the least of her talents. Throughout her education, high marks filled her transcript with only minimal effort. She won awards as a writer, she earned applause as a quick-witted public speaker, and she was being singled out for the chair of an international business association. Ambition filled her veins, self-reliance and independence exuded from her smile. Of course the guys lined up; she had sixty-five flower arrangements to take home after her hospital stay.

Melissa brought the same aggressive, competitive energy she used in sports to the boardroom table. As director of public relations for Lakehead Psychiatric Hospital, Melissa's vivacious appeal mirrored the image the hospital wanted to present to the surrounding community. She turbocharged the organization with new, socially minded initiatives, such as the provincial mental health strategy. Doctors turned shy in her presence and coworkers gossiped about her availability. In every instance, she exemplified the exceptional. At just thirty-one years of age, Melissa Felteau felt the adoration and envy of the world around her. And she enjoyed it, perhaps a little too much.

The truth of her postinjury life unveiled itself in the most casual heartbreaks. Each day served up a reminder of skills she had lost. At first, her sisters commented that she sounded strange on the telephone, when, in fact, she had developed a stutter. After Melissa returned home under her parents' care, she moved about as though sedated, even after the pain medications wore off. She couldn't shake the slumber. Her vision was plagued with floaters. Coffee mugs and books fumbled from her uncoordinated fingers. She tumbled from table to bed to sofa and back again. After days of tripping over herself, she realized that she couldn't seem to regain her balance.

Canada's healthcare system allotted for an unhurried two months' convalescence, an interminable period of rest for Melissa. She yearned to return to work, to get back on track with her life. She ignored any cautions from doctors and family as unhealthy pessimism, hardly noticing that fewer friends were calling or visiting. The first day at work, however, was a distressing confrontation with reality. At first, she could not figure out which of her keys opened her office. They all looked the same to her. She then tried to read a memo on her desk and found that she couldn't understand a simple paragraph. The hum of her computer drove her crazy, and the ringing in her ears wouldn't go away. When she attempted to visit coworkers, she lost herself in the once familiar hallways. After she found the right department, she couldn't remember the name of the person she intended to visit.

"Names are very important to me," she says. "You can imagine how embarrassing it was for me to be the public relations director and not remember names."

Melissa endured two weeks of work before she requested a vacation. The additional time off didn't do the trick. When Melissa resumed her job, she once more attempted to pick up the rhythm she had sustained for the previous six years. A presentation that normally would have taken her an afternoon or two required all of her focus for weeks. Without realizing that her executive functioning was impaired, she struggled with duties involving multitasking, decision making, and organizing. There were awkward moments when she cried at the slightest provocation or snapped at her assistant. She had once excelled at complicated managerial tasks; now the most menial chore, such as writing a complete sentence, proved a challenge.

"I was overwhelmed," Melissa confesses. "I was having major problems and didn't want people to think I was stupid, so I stopped hanging around people from work."

Social engagements became opportunities for embarrassment and ridicule, causing Melissa terrible personal conflicts. She wanted to be out among the crowds, but simultaneously felt vulnerable and frightened by them. Melissa sank into long sulks and quiet withdrawals. The invitations stopped coming and the phone rarely rang. Of the sixty-five well-wishers, only one person remained a friend.

Melissa exchanged dates with doctors for appointments with doctors. On a visit to one doctor's office for her pain issues, she happened across some literature on brain injury that had been left in the lobby. The brochure listed the wide-ranging side effects of brain injury: mood lability, agitation, poor attention, memory problems, coordination difficulties, and disorientation.

"I saw it and realized, 'Oh my god, this is what I've got,'" she says.

Once I've finished my evaluation of a brain injury survivor, after I've put my file back in my case and put away the pen, I often wait around until I sense they're ready for me to leave. They'll ask me about my job, comment that I look tired, invite me to stay for dinner, show me pictures, ask if I have any pictures of my baby. It's the best part of my job. It's in that space, away from the clinical microscope, that the damaged blossom. They will admit that they haven't felt like their old selves, or they will confess, with a little shame, about how they never paid any attention to disabled people until now, and look at me, I can't believe I am one of them. "I'm not me anymore," one survivor told me, "but I'm still me." I take those statements as an indication of deeper musings. Every so often a survivor, speaking to themselves more than to me, will ask the most spiritual question.

The most spiritual question in the world is not whether there is a god, or how we came to be in the universe. The most spiritual question in the world does not concern itself with knowing why there is suffering or why we are here; those ponderings stem from the most spiritual question. The aim of every mystical tradition in any religion is a sincere and relentless pursuit of the answer to the most spiritual question. The most spiritual question is about you. The question is: *Who am I, really?* Brain injury, above all other anguishes known to man, perpetually invites us to embark on the search for our selves. Who are we, other than our brains, really?

Different spiritual practices offer various methods for inner investigations, but few religious traditions are so doggedly concerned with the matter of self-discovery as Zen Buddhism. If a person is able to clearly realize their true nature, the thinking goes, then they will experience a freeness of being that acts as a boundless source of strength. One glimpse of your true nature is enough to dislodge former perceptions of your self, and you begin acting from your new understanding.

After the initial confrontation with their true selves, certain Zen students begin a rigorous curriculum of enlightenment that begins with the study of one of the world's most enigmatic and profound sacred texts, *The Gateless Gate*.[1] Compiled in the thirteenth century by the Chinese monk Mumon, the writings contain dozens of koans, any one of which could take years for a Zen practitioner to complete. The koans are brief, sometimes irrational vignettes that relay encounters between monks and Zen masters. In the exchanges, some monks attain enlightenment, while others simply go on about their business. Any koan has the potential to trigger a direct experience of the sublime, a deeper realization of the self.

Beat culture popularized some of *The Gateless Gate*'s koans, such as "Does a dog have Buddha nature?" or "What is the Buddha? (answer: three pounds of flax)." Though they're publicly received as absurd and ridiculous parables, to the Zen student koans are keys to unlocking truths not otherwise accessible. Taken individually, koans have a subversive effect on the rational mind; part of their power is in their ability to confuse and deflate any attempts at intellectualization. A Zen master might put his shoes on his head, or simply advise a monk to wash his bowl. Viewed collectively, however, the koans yield other rich subtleties.

The Japanese phrase *chinami ni*, for example, appears in fourteen of the forty-eight koans. Contextually translated as "in all earnestness" or simply "earnestly," the phrase describes the attitude of a monk who is seeking illumination. A monk will approach a teacher in all earnestness, a question will be posed earnestly, a gesture is conducted in all earnestness. As each koan encounter relays a moment ripe with the potential for enlightenment, the virtue of earnestness is regularly dealt a passing reference. Earnestness, the sacred text implies, is a prerequisite for revelation.

Six months after the injury, Melissa arranged for her first neuropsychological evaluation. The testing took several hours to administer and consisted of the Halstead-Reitan neuropsychological battery, along with ancillary measures such as the Weschler scale, the Wide Range Achievement Test, the Peabody Individual Achievement Test, the Detroit Tests of Learning Aptitude, the Rey-Osterrieth Complex Figure Test, the California Verbal Learning Test, and the Minnesota Multiphasic Personality Inventory. The testing left her feeling depleted. At least she would finally have a solid idea about what kind of problems she was having, she thought. She could not have anticipated the actual results:

> Lengthy or complex auditory/verbal information is assimilated with difficulty, with relatively greater decrements in recall demonstrated for this type of information after delay. Significant difficulty is demonstrated in

> the initial assimilation, processing, retention, and later recall of nonverbal information . . . A general slowness in the efficiency and processing of information is demonstrated by an individual who also encounters difficulty with more complex types of cognitive tasks . . . There is no doubt, however, that given this individual's industrious and hard-driving style that compromised mentation and an inability to execute her normal occupational responsibilities and social obligations have tended to exert a profound and negative impact upon her psychological and emotional functioning at this time.

As Melissa flipped back and forth through the thirteen pages of the report, the words struck her like a litany of shortcomings. Even in areas she considered innate strengths, such as verbal ability and comprehension, she now appeared average or slow. She cringed at the psychological impact demonstrated in the findings, sensing that they were more accurate than she wanted to admit. The neuropsychologist had identified over a dozen general areas of cognitive impairment and detailed more than forty distinct impairments. In his report, he suggested follow-up appointments at the Organization for the Multi-Disabled, a crushing blow to a woman who so recently considered herself extraordinarily able. According to the conclusion on the report, there was no longer anything exceptional about Melissa Felteau.

For six years after the evaluation, Melissa endured all the same frustrations and setbacks common to brain injury survivors. The neuropsychologist had recommended a specific course of treatment, but in the end, the healthcare system in Canada was no better equipped to handle brain injury than that of any other country. She struggled to find service providers and was forced to travel hundreds of miles to see specialists who could do very little for her. One occupational therapist came into her home and offered to organize her closet by color, so that she could pick her clothes out more efficiently. Melissa felt humiliated by the therapist's perception of her. She didn't need help getting dressed; she needed her old brain back.

Because of the continuing costs of her care, her case mushroomed into a long-standing litigation with her insurer, which brought her more stress and health problems. As a particularly malicious tactic often used in similar cases, the insurance company claimed they had overpaid her, so that she would be put into financial duress. The tactic worked. Her financial life teetered on the brink of bankruptcy; even her parents suffered setbacks to their retirement on her behalf. She eventually resigned from her job, feeling conquered and inadequate because she simply could not meet her responsibilities.

Physically, Melissa had deteriorated significantly. She ached to return to a swimming pool or lake, but when she entered the water the motion caused her to lose her bearings to the point where she couldn't figure out which direction she faced. Swimming felt like drowning. As if the prison of inactivity wasn't restrictive enough, Melissa developed regular bouts of idiopathic anaphylaxis, a severe, stress-related allergic reaction that required her to be within forty-five minutes of a hospital at all times. Almost twice a year, Melissa's body would revolt with anaphylactic shock. Her throat would constrict until it closed, her eyes would swell shut, and hives covered her body until a strong-enough steroid could be administered in an emergency room. She was well aware that the reaction could someday bring with it another brain injury, or possibly death.

Melissa might have been rendered helpless against her brain injury, but she felt certain she could find a way to minimize the anaphylaxis. She began to research stress reduction techniques and eventually stumbled onto a holistic center that offered a five-day integrative program based on techniques developed by the University of Massachusetts Medical School. The core principle of the program was the cultivation of mindfulness through guided meditations. As Melissa studied the program in detail, she worried that her brain injury might not allow her to fully benefit from the program. She decided to enroll anyway.

"Those five days were grueling," she recalls. "We would start with gentle yoga, then meditate until nine at night." During that retreat Melissa learned about the various attitudes of mindfulness, and she participated in self-discovery exercises. The point, as she understood it, was to tap into the powerfully curative characteristic of awareness and to generate so much of it that she might be able to apply it to ordinary everyday life. Even the most mundane act, like washing dishes, could be transformed by simply paying attention.

It was during one of the exercises that Melissa experienced a sudden clarity. She had been instructed to think about the pleasant events in her life, and she noticed that each time she did, she would undermine herself with a barrage of malicious thoughts.

"That was my biggest problem," she tells me. "I realized that I was always comparing myself to my preinjury self. I was trapped in a vicious cycle of rumination and depression."

That one instant of awareness shed enough light to convince her to continue in her development of mindfulness. Maybe this was something she could be good at, at last. After the retreat, Melissa began a disciplined schedule of daily meditation and yoga. Although her initial intent was to alleviate the likelihood of anaphylactic shock, she began to notice other changes.

"Right away it improved my concentration, and it improved my mood swings," she says. "I was a lot less irritable, and my memory started improving because I could attend more. My family was even starting to notice all this."

During a trip with her parents and her sister, Melissa began to see the signs herself. Prior to meditating, she would get sickened just watching the trees whip past the car window. Typically, one day of family interaction exposed her to so much stimulation that she needed to withdraw out of exhaustion. On this trip, however, she stopped cocooning herself in her cabin room. The headaches and dizziness had evaporated, along with the agitation and moodiness. When Melissa emerged from her room the following day and joined her family, they were shocked at her presence.

"You can be around us again!" her sister said.

The changes were so significant that Melissa wondered if mindfulness could benefit other brain injury survivors. Knowing that graduate school might prove an insurmountable hurdle, she applied and was accepted into a master's program in adult and continuing education. Through her affiliation with a local university, the encouragement of her professors, and support from a hospital, Melissa made an arrangement to conduct her first foray into research. She had chosen a group of brain injury survivors who agreed to engage in a seven-week training in mindfulness. Melissa would then track the group for any indicators that the practice changed their lives.

. . .

Case 19 in *The Gateless Gate* contains the following koan:

> Joshū earnestly asked Nansen, "What is the Way?" Nansen answered, "The ordinary mind is the Way." Joshū asked, "Should I direct myself toward it or not?" Nansen said, "If you try to turn toward it, you go against it." Joshū asked, "If I do not try to turn toward it, how can I know it is the Way?" Nansen answered, "The Way does not belong to knowing or not-knowing. Knowing is delusion, not-knowing is blank consciousness. When you have really reached the true Way beyond all doubt, you will find it vast and boundless as the great empty firmament. How can it be talked about on a level of right and wrong?" At these words, Joshū was suddenly enlightened.

Mumon, the compiler of the koans, included a brief commentary and a verse with each koan. The verse that accompanies Case 19 offers a delicate hint toward unlocking the koan's peculiar teaching.

> *The spring flowers, the moon in autumn,*
> *The cool breezes of summer, the winter's snow;*
> *If idle concerns do not cloud the mind,*
> *This is man's happiest season.*

Although Case 19 begins simply enough, it appears to fall back on itself with contradictions, pitting action against inaction, the dual against the non-dual. For contemporary readers, though, the koan presents a more literal complexity. What, exactly, is meant by "mind?" The most commonly held belief among neuroscientists is that "mind" refers to the by-product of brain; it is the whir of our cognitive wheels spinning. In academic journals and the common vernacular, the terms "mind," "consciousness," "ego," "soul," and "awareness," are all used interchangeably, as though they refer to the same process.

Tibetans, the great investigators of the inner universe, delineate between multiple types of consciousness—an acceptable number includes about eight kinds in various Eastern philosophies. For decades, our researchers have bemoaned the woeful ambiguity of the terms "mind" and "consciousness," and they note its tendency to obfuscate the outcomes of serious neuroscience.[2] Western science has yet to offer a standardized definition. It is at this semantic junction that we can most pointedly feel the shocking depths of what little we actually know about the brain. We can place shunts in the middle of someone's head and we can shock depression out of the skull, but we don't know the foggiest thing about ordinary mind. Science can tell you the molecular composition of each neurotransmitter suspended in your brain, but it cannot tell you who you are.

Melissa drives me to a large hospital where she's conducting her third study on the effects of mindfulness-based therapy on brain injury survivors. The first two studies yielded such impressive results in areas such as quality of life and the relief of depression that Melissa received a grant from the Ontario Neurotrauma Foundation to continue her research. Although she can manage only two treatment arms in her current location, she already has thirteen hospitals throughout Canada eager to participate in the study.

The research room contains four rectangular folding tables arranged in a large square, and by the time I arrive eight of the participants are already seated. The participants range from nineteen years of age to sixty, and none of their injuries are immediately evident, though several of them wear head coverings. I look around the room and see no wires or electrodes, no fancy sensors or expensive scanners. There are simply eight humans, each with a damaged brain, and their faces are serious. For them, this could be the most important change in their lives. Melissa introduces me to the group, and I smile at everyone and wave a hand, and beg them to ignore me altogether. I just want to observe, I explain, and they return nods and smiles and are gracious enough to let me stay.

Melissa opens the group with a brief review of the exercises they completed last week, and she calls for completed "homework" assignments to be handed in. As part of their responsibilities, the participants agreed that they would dutifully complete assignments given to them. Without exception, each one turns in a form to Melissa that charts the duration of their meditations and the types of mindfulness exercises they conducted at home. Melissa then announces that she'll begin with the first mindfulness exercise, and all around the room the participants sit upright on the edges of their chairs, each of them poised but relaxed. I decide to join them in the exercise, so I set down my notes and lay my hands in my lap.

Melissa begins talking in a low, calm tone, and she encourages the group to pay attention to their bodies. She tells us to intentionally relax the tensed areas and to notice any painful areas. She directs us to breathe deeply, and her voice, now slow and patient, continues to massage the room. After several minutes of stillness, she resumes her talk and encourages us to pay attention to our thoughts. It's okay to have them, she says, but it's not okay to feel bad about them. Just watch them without judging, she says. I rebel and peek around the room. Everyone's eyes are lowered, and their faces are slack. They look as if they've had a long rest already, and we've only just begun.

The script Melissa follows comes from the same series of guided meditations she first encountered at the retreat center. In various ways, they encourage and exercise the different tenets of mindfulness, from patience and trust to acceptance and letting go. Participants are then taught how to apply their new insights toward their personal lives. Melissa's hope is that everyone in her group is able to find some respite from the many challenges that plague them daily. Once this particular eight-week group has concluded, she'll continue to follow up the research through questionnaires in order to test the durability of her experiment.

Mindfulness has its effect on us. It takes about forty-five minutes for Melissa to narrate the entire meditation, which is no small achievement for several people in the room who struggle with attention deficits. One woman breaks into tears at the end of the exercise, crying out that she's not sure if she's doing it right. A younger man confesses that he almost fell asleep during the talk, which was strange for him, because he has problems sleeping. Another woman tells the group how her anxieties are finally starting to let up, and the group nods, as though this is a common sentiment among them. They continue to process their experience for a few minutes, and then Melissa gives them their assignments for the following week and encourages them to contact her with any questions that arise later. Some of them stop to thank her on the way out, and soon they are all gone. From the stillness I draw a curiosity, and I approach Melissa to ask her how mindfulness works.

"It gives you a new way of looking at life," she tells me. "This type of therapy teaches you how to stop wanting something different than the life you have, and there is incredible freedom in that. We come to realize we are whole, no matter the deficits, and that there is more right with us than there is wrong. To really embrace that is transformative."

I tell Melissa that it sounds like a wonderful and noble pursuit and that I hope her research opens up new solutions for survivors in the future. Then, being unable to resist, I ask her what mindfulness training has done to her sense of self.

"Who are you, really?" I ask.

"I love myself the way I am now," she says. "I appreciate that I am not my brain injury. It was a traumatic experience, to be sure, but it deepened my relationship to myself and to others. I have become a more loving person. I am a lot more empathetic, and I know what compassion is now."

The neuropsychological evaluation portrayed a woman about whom nothing seemed special, and in a sense, there is nothing special about Melissa Felteau anymore. She gets a little confused with directions, and she still goes off topic in conversation. She becomes a little flustered when she tries to parallel park, and she laughs at herself when she trips over a rise in the sidewalk. She is a student conducting her research, and she is a daughter who loves her parents. The spring flowers, the moon in autumn. Melissa is brain injured and she is wonderfully whole. Using only her ordinary mind, she has found her way.

THE HOSPITAL IN THE DESERT

A year ago, an explosion disassembled Private Juan Morales, and some of his missing body parts are still on back order. He slurs a "hello" when he's introduced to me, then bitches a little about cafeteria food at the Minneapolis Polytrauma Rehabilitation Center, where other Iraq War veterans receive care. An entire quadrant of Morales's head is gone, pending a cranioplasty, and his scalp is a mottled patch of scarred-over shrapnel divots, nickel-size tufts of black hair, and miscolored skin grafts. Shortly after the blast, a microscopic peek into his brain would have revealed a gray mush of tangled axons and dendrites. He shouldn't be able to do much more than drool, roll his eyes, and maybe grunt, but he's smirking and gesturing with ease.

Morales, along with the rest of Iraq's wounded, are redefining severity. A hundred years ago, a horse's kick to the head would have done you in. In Iraq, you can take a golf ball–size missile through the skull and survive. Thirty years ago, a severe brain injury meant that you couldn't go to work anymore, and five years ago, a severe brain injury could mean a long coma, a minimally conscious state, or a daily condition of total dependence. In the regrettable math of the Iraq War, a truly severe brain injury now comes with polytrauma: brain injury plus a plastic arm, minus your balls, divided by weeping burns. You can survive more injuries than you'd care to know.

In a few more months, Morales will complete his trip through the military's medical megaride. He'll get a new skull plate, some more plastic surgery, hair transplants, and a prosthetic arm and leg. Time will buff away the Frankensteinish seams. He should roll out of Minneapolis shiny and finished, looking as though the blast only tempered him for the future ahead. The missing limbs won't deter Morales much. It's the injury you won't see, the blasted brain, that won't be rewired so beautifully. Odds are that it will undo every aspiration Morales ever harbored and force him to adopt an entirely unpredictable future, one that thousands of brain-injured servicemembers are facing. At least Morales will have aspirations, hopes. I can't say the same for most injured Iraqis.

Above the entrance to the Air Force Theater Hospital (AFTH) at Balad Air Base hangs a simple sign with red letters: EMERGENCY DEPT. For more than eight hundred people a month—a quarter of them Iraqis—admission into

Balad Hospital signals a turning point in their lives, a moment in time in which their future takes a direction they had never planned or wanted. If you're an American, military or contractor, Balad is the gateway to an elaborate, frenzied ride through the most aggressive medical system in the world. Wounded Iraqis, however, go through the experience on an altogether different plane; for them, Balad Hospital is the seat where God wields his will.

Among its many honors, Balad Hospital has an unholy distinction: it is the brain injury capital of the world. The hospital at Balad houses the only head and neck surgical team in Iraq, so it sees more damaged brains pass through its doors than any other facility in the world. After a few months of practice at Balad, neurosurgeons are emerging as the most experienced neurotrauma specialists in history.

While on business in Alaska, I had the opportunity to visit with Dr. Eli Powell, a former commander of AFTH Balad, while he was stationed at Elmendorf Air Force Base. Powell has kind eyes, salt-and-pepper hair, and a gentle voice. He looks like everyone's older brother or favorite uncle, but after a few minutes of speaking with him, I felt as if his skull housed two brains. He was effortlessly reading and responding to e-mails while talking to me, and he never showed any indication of distraction when answering my questions. I kept egging Powell for details about the hospital, about the Air Force's well-publicized flying intensive-care units, and the consequent treatment of brain-injured troops at Landstuhl and Walter Reed.

"Why don't we just send you on a mission over there?" Powell asked. "I could set it up for you, and that way you can see it all for yourself."

I couldn't decline. I had looked Private Morales in the eye and seen the future of brain injury, and I knew more like him were making their way back into their communities. The global war on terror has already yielded more than ten thousand survivable traumatic brain injuries to American troops,[1] and there are no indications that the rate will slow. Balad Hospital proves that we are no longer asking most soldiers to die in service; we are asking them to accept a lifetime of severe disability. It's a big sacrifice, one that's nearly impossible for a new enlistee to grasp.

Should a brain injury befall you in America, you stand a 71 percent chance of being alive one month after your ER visit. If a brain injury occurs anywhere in Iraq and you're medevaced to Balad, your chances of survival skyrocket to 98 percent, the highest rate of survival for any trauma hospital in history. What takes the average emergency room several hours to accomplish, Balad orchestrates in minutes. Record time from admit to operating room is eighteen minutes—and that includes CAT scan and lab work. Twenty years from now, American trauma care will be modeled after Balad Hospital's pioneering work.[2] It is healthcare unencumbered by insurers and accreditations.

"It is the purest medical mission in the world," Powell explained to me, "but it's not one you would wish on people."

I wasn't able to appreciate Powell's comment until I visited the hospital in the desert and saw its inner workings myself. I now know the kinds of things you wouldn't wish on people.

The descent into Balad is a tangle of bureaucratic, logistical, and psychological thorns. Although I had already been vetted as a brain injury case manager, it took me eight months of screenings and cancellations in order to gain clearance, and another two days' worth of travel just to get onto the Air Force base in Ramstein, Germany, where most flights into Iraq originate. Everyone on the flight wears a long face that only grows longer with the passing hours. Hardly anyone moves around the hull of the C-17 until just before the descent into Iraq, when the lights inside the craft change from pale green to apocalypse red. We grudgingly strap on our bulletproof vests and Kevlar helmets, and a voice on loudspeaker instructs us to strap our asses down tight. Everyone does. Moments later, the nose of the plane lurches downward, a combat landing maneuver, and we fall headlong into the Iraqi night.

Descending into Balad, I know that I will be seeing brains, I just don't know whose. I imagine it will be a soldier, maybe a Marine. I envision the surgeon sawing away half the skull and the bone flap lifting dramatically, as though it had hinges like a jewelry box. I will see a brain, it won't belong to a Marine, and it won't open that way at all. It will belong to seven-year-old Jassim, who should have been sleeping alongside his little sister this night, dreaming, while I am soaring miles above him. His brain will open like a hard-boiled egg opens, with a tap, a crunch, a starburst crack.

When the cargo hatch opens, a hot, industrial odor permeates the craft; the runway lies downwind from the base's landfill, which smolders like an active volcano around the clock. Covered in an industrial night smog, the base looks like little more than a landing strip with a few distant hangar pods. Not long after deboarding, I am ushered to a trailer fortified with sandbags. The small bare room houses two twin beds, a nightstand, a lamp, and an electrical outlet. My media escort for the trip informs me that these are the quarters typically reserved for dignitaries.

In daylight, Balad Air Base resembles a construction site; everything you see seems made of concrete or steel. The base, nicknamed Mortaritaville, receives attacks every day, a circumstance which has sprouted thousands of reinforced T-barriers that surround every facility. Every intersection looks like the same convergence of sand, gravel, and concrete. Even the water doesn't know where to go. There isn't a drainage system, and the earth is packed so tight that a simple rain floods the grounds, turning the desert floor into a rust-colored soup of mud and stone. Soldiers clomp through puddles, and the puddles seep into places you don't expect to see them, like hallways and intensive care units.

Located near the Tigris and occupying lands once farmed by sky-worshipping Sumerians, Balad Hospital hallows the earth below it with its steady flow of sacrificed blood. And like the tabernacle of the Israelites, located five hundred miles to the west and two and a half millennia in the past, all of its mysteries are enshrouded in tents. The Ark of the Covenant was veiled in a tent within a tent; AFTH Balad is a series of more than thirty-five thousand square feet of connected tents. Balad's Holy of Holies, the six operating tables, are housed in three retrofitted shipping containers.

I arrive through the back door of the hospital on Man-Love Thursday, the day of the week when the majority of male clinicians wear floral print scrubs, just to, you know, change the pace. With surgeons submerged in the rhythm of suture and saw for days on end, any deviation is welcomed, no matter how emasculating. The hospital's deputy commander, Colonel Lorrie Cappellino, gives me a tour of the facility. Although Cappellino has a motherly manner, she sports desert camouflage and packs a pistol. She explains that the hospital's admit rate is dictated by insurgency, but that she coordinates the flow. Only the Iraqis slow the pace.

"You'll notice that all the Iraqi patients have Ugandan security guards assigned to them," Cappellino tells me as we pass one of the guards holding a rifle. "Some of them are insurgents and we don't want them roaming the hospital unattended."

Two parallel hallways, each about a hundred yards long, form the hospital's main track, from which nearly all the tents connect. Near the back end of the hospital are the tents that house administration, food service, and communication. Along one hallway runs the intensive care units, occupying three tents, eighteen beds total. Iraqi patients occupy two of the three units.

"We have fifty-eight beds total," Cappellino tells me as we march. "In a pinch we can surge to a hundred."

Nearly every exposed piece of framing is plastered with good-job cards from American kids, or with permanent marker propaganda, pro- and antiwar. The informality of canvas ekes into the psychology of the hospital, investing it with an organicity undetectable through television sets and movie screens. With rainwater slicking the hallways, mud tracks throughout the building, and the subtle billowing of tents as they respond to the climate, the hospital feels earthy, worn, and human.

The physicians' lounge at Balad, for example, is located in a tent attached to the emergency room. There are four Formica desks along one wall, and two abused faux-leather couches along the other. A sleeping doc on a couch is a common sight in the room, even though the hospital's main corridor runs right through it. In the same lounge, above the trauma chief's desk, hangs a whiteboard where the doctors capture one another's asides:

> If I had Iraqi feet, I'd be Barney Rubble.
> I've been to Africa, and it ain't this hot.
> If I lose both my balls, you can take my eyes out.

The doctors work on patients, the hospital works on doctors. Medics don't realize their thresholds until they return to their business park offices and impatiently listen to complaints about tennis elbow and sore throats. Balad bestows skills they'll never need at home, and a few they never wanted. All the medics can distinguish an approaching Black Hawk from other helicopters by sound pattern alone. They sense trauma coming before the announcement on loudspeaker, and they're comfortable not knowing who or what kind of injury awaits their attention. Patients get wheeled into the ER tent bearing gruesome secrets: blown-open sternums, prolapsed intestines, bleeding eardrums. It's up to a dozen clinicians to assess and stabilize each patient before a new one appears. A new one always appears.

Balad's Emergency Room is striking for what it lacks. There isn't a waiting room, and there isn't a need. I'm not in the hospital long before Jassim is rushed in on a rickshaw gurney. Swaths of white gauze circle his head, exposing only his eyes, mouth, and nose. His face is a dark purple bruise mashed with dried blood and dirt. Jassim's eyes are closed and his limbs flaccid, and he doesn't react when the trauma specialists push needles into his arms. He is immediately surrounded by a quiet swarm of trauma specialists.

Heartbeat, slow. Breathing, weak. It takes but a moment before the trauma team decides to intubate Jassim, and the tube is pushed down his throat. At the same time the anesthesiologist gives him an antigagging sedative. The drug is too slow, though, and Jassim's flimsy hand reaches up and swipes at the ventilator tube, and everybody smiles. The oncologist at the head of the gurney catches Jassim's hand and it drops weakly, then swipes again, and drops. On Jassim's third try, his hand lifts a few inches and collapses. It is the last time I will see Jassim move.

Cappellino appears beside me and, in a whisper, points out the various clinicians. "We have a board-certified emergency room physician doing triage on the patient right now," she tells me. "I can see pharmacy is here, there's a trauma physician tapping at his belly, this is another ER physician inspecting his head. We have another trauma specialist at one end, an oncology specialist there, a respiratory technician, an army medic, a medical technician starting lines for IVs—you can see three bags hanging; they're going to start giving blood."

The trauma specialists, more than a dozen of them, have done this hundreds of times already, reducing almost all need for verbal communication. They move wordlessly, nodding, hands weaving up and down the body, while clinicians orbit the gurney and one another. It's a ballet of blood pressure pumps and IV bags. Within six minutes, they have Jassim fully assessed, X-rayed, and on his way to the CAT scanner. The on-site lab will process all the blood work.

For fear of retribution on Jassim and his surviving family members, the bombing's details must remain a secret. The generalities involve a suicide bomber and a mosque, and an explosion that killed dozens and wounded many times more. Jassim, not old enough to be in third grade, is among the living, but barely. His sister and some classmates are among the dead.

There are two routes into Balad Hospital. You can go through the back door, as I did, or you can arrive by helicopter. The wounded are immediately swept off the helipad and carried through a canvas canopy called Hero's Highway. The first thing a patient sees, if they are conscious, and if they still have eyes, is a massive American flag attached to the corridor's ceiling. The flag is supposed to communicate assurance and instill courage, but its actual message only adds to the emotional conflict felt by servicemembers.

When I first see Sergeant Raleigh Heekin in the ER, he's lying on a stretcher, unattended, with his patient number scrawled on his bare chest in permanent marker. I ask him if he can remember Hero's Highway.

"When I saw the flag it made me feel good," says Heekin. "But right now, I lost two very good men, and that doesn't . . ." Heekin stops himself midsentence. "I lost two guys today. Now I'm questioning everything, whether I should be in the army, or get out. I don't know what I'm going to do from here."

That morning Heekin's vehicle hit an improvised explosive device, an IED. The blast degloved the underside of his left knee, tearing open his leg from calf to midthigh. He awoke from the blast to find himself covered with his medic's insides.

Heekin's leg will get more attention than his brain at this point, even though I can hear the brain injury as he tells me his story out of sequence. One of the most frightening things about blast-related brain injuries is that they can pack up to four assaults in one instant. The first few milliseconds bring atmospheric fluctuations that expand and contract the brain. Skull-penetrating shrapnel and debris follow. The head invariably slams into a nearby object, and lastly, the brain releases a cascade of neurochemicals that cause further injury. Nobody really knows what is happening to the brain at a cellular level during a blast injury, but the aftermath is spread diffusely throughout the tissue, resulting in microscopic tears and hemorrhages.

Heekin doesn't have a penetrating injury, but he describes losing consciousness several times following the explosion, with some of the episodes lasting minutes.

"After I got up out of there, I just had to sit down," he explains, trying to make sense of his lapses. "My mind just gave up."

Had Heekin not received a full dose of morphine to alleviate the leg pain, he may have been able to receive the Military Acute Concussion Evaluation (MACE) at Balad, a twenty-minute neuropsychological test that the military now gives to every American wounded in a blast. The test has only now become compulsory; no similar evaluation occurred for the first four years of the conflict, contributing to the speculation of high numbers of undiagnosed injuries. While the MACE may someday be administered beside football fields and boxing rings, it has one significant limitation. The test does not cross the cultural divide; Iraqis don't receive it.

The flag at Hero's Highway carries a different connotation for the thousands of Iraqi police and civilians who are also treated at Balad Hospital. They know, through word of mouth, that they'll receive excellent care, but they also know that their admission marks them in their community. They and their family members could be killed for receiving American help.

"Once they are here, most Iraqis are very afraid to leave the hospital," one of the translators at the hospital told me. "For them, it is a death ticket."

Doctors anxious for a breath of fresh jet exhaust sometimes steal up to the roof of a nearby storage building for a round of cigars and swearing. It is on this rooftop that I am able to join Dr. Richard Teff, one of Balad's two neurotrauma surgeons. Teff has short dark hair, wears glasses, and, even with cigar in hand, gestures with a surgeon's economy. When he speaks, his voice is direct and carries reassuring tones of disillusionment, confidence, and anger. Teff tells me that despite the irregular sleep, long hours, and four-month rotations, the Balad medics have it much easier than their Iraqi counterparts. I have heard that Iraq is facing a crisis of fleeing doctors, so I ask him what he knows about the numbers.

"I can tell you that at Medical City hospital [an Iraqi hospital in Baghdad] that I have been to in the past, they used to have five or six staff neurosurgeons," he said. "All the consultants there were either killed or left. They have just one consultant now."

"I hear they fetch a pretty high price in terms of kidnapping," I say, citing a recent ransom figure of a quarter-million dollars.

"Absolutely," he says. "There's a guy that I know, his name is Khalili, who was doing great things for Medical City, and while he was down there, he was held for ransom once, his car was stolen at gunpoint once, his family was threatened, and he finally moved to Canada. Most of the consultants that are there now live in the hospital. I spoke with a vascular surgeon two weeks ago, and he said that he hadn't seen his family in eighteen months because he hadn't left the grounds. It is a sad state of affairs, but if the bad guys find out a doctor is married to somebody, they will hunt down their family and kill them. And if they know that someone works in any sort of capacity assisting the Americans, then they'll hunt them down and kill them, too. Most of those guys did what I would do: get the hell out of there."

Teff's cohort is Dr. Mark Melton, the other neurotrauma surgeon at Balad. Melton is tall, with a boyish face and manner, and acts demure when pressed about his job. He claims trauma surgery isn't as exciting as tumor removal, and he seems gleeful that he got to clip an aneurysm in the operating room about a month ago. When I join Melton and Teff for their patient rounds, we approach the bedside of Adila, an Iraqi girl. She's in one of Balad's three intensive care units, two of which house only Iraqis.

Adila isn't supposed to be here; the Air Force Theater Hospital at Balad Air Base isn't a pediatric hospital, although 10 percent of their patients are children. A track of fifty staples trails the length of her torso; fresh stitches—vestiges of an ulnar nerve repair—circle the circumference of her left elbow. Both of her eyes were decimated in the mortar attack and consequently removed. Adila is two years old, and she has little bits of shrapnel lodged in her brain; extracting them would only risk more damage. She's thrashing around in discomfort, and it brings a bittersweet smile to Melton's lips.

"That's a good sign," he explains. "She calms down when you pick her up and hold her."

Adila's ability to respond to pain puts her in a category above the other brain-injured Iraqis who lie beside her, several of them occupying ventilators. No response to pain would lower her GCS (Glasgow Coma Scale) score, which could mean a quick move to palliative care, a final act of humanity before the morgue.

"You have to keep in mind what is going to happen with the patient once they leave the hospital," Teff says. "If you go to an Iraqi hospital, chances are high that you're not going to have a neurosurgeon there, and the general surgeon can probably clean things up but they are not going to go inside the head and do something they don't feel comfortable doing. What they do is just take them home and pray, and if it is God's will, the patient will do well. Most of the time the patients die."

When I look at Adila, I can hardly take the confusion in her face. She's in a great deal of pain, she's crying, and she has no idea how her little life amounted to this. There was a bang, and the light of the world vanished. She doesn't recognize anyone's voice, and English words don't make sense. The pain in her head has to be an unquenchable fire, and the incisions and puncture wounds across her torso and limbs sting every time she moves. What does she know of nation building and religious law? She calms down when you hold her.

Adila's parents do come around to see her, Melton later explained. The problem is that she's now considered damaged goods. She is blind, has heart problems, and is being fed through a tube. At this point in Iraqi society, a disabled girl is a major liability, and the girl will most likely be discharged to Medical City; Adila or her parents may disappear in the process. If it is God's will, she will do well.

. . .

As a liaison caught between the military and the Iraqi Ministry of Health, Department of Defense contractor Dr. Adam Abraham feels that he is doing as much as one person can manage. When medical supplies expire at military hospitals in Iraq, he has them transported to Medical City. Although half of the supplies are often looted before they arrive, clinicians in Baghdad feel the expired goods are a godsend. If Medical City has it so bad, I ask Abraham, what are the other hospitals like in neighboring cities?

"I would not take my dog for treatment in some Iraqi hospitals," says Abraham. "You can smell the infection from the hallways."

Curious, I ask Abraham about the history of the Iraqi healthcare system.

"In the late eighties and early nineties, Baghdad was the city of choice for Middle Eastern medical care," he explains. "Now I would not give their medical graduates the equivalent of the nursing degrees in the U.S."

Both Abraham and Teff underscore the absence of services we consider standard, such as nursing home care and rehabilitation.

"They don't have any therapists: physical therapists, occupational therapists, speech rehabilitation, cognitive therapy. That's not available in Baghdad at all," says Teff. "I don't think the Iraqi surgeons are particularly aggressive about treating head injuries. It's much like it was for us during the Civil War."

Teff isn't exaggerating. Without the reliable use of CAT scanners, MRIs, and surgical microscopes, Iraqi neurotrauma surgery amounts to peering directly into the head and pulling out whatever doesn't appear to belong. Moreover, needles and blades go through multiple uses. Painkillers are a luxury first afforded to the operating room and rationed thereafter. Anesthesiologists routinely cancel surgeries due to lack of oxygen. Screams and cries set the milieu, night and day.

"There is an ongoing state of controlled anarchy there," says Teff. "A process of damage control care that provides the basics to those who need them worst, and palliates those who are unlikely to recover quickly."

Dr. Melton is at the computer when the results of Jassim's CAT scan arrive a few minutes after his admission. While Melton eyes the screen, he tells me to go ahead and get on scrubs to meet him in the operating room. He points at a subdural hematoma creeping through Jassim's frontal lobes, and explains that he's unsure about Jassim's cranial integrity.

"There's no humongous skull fracture that I can see. But that," says Melton, pointing near the top of Jassim's skull, "that looks disconcerting."

While Jassim is being prepped for his operation, Sergeant Heekin embarks on a journey of a different type, a journey Jassim won't make. Heekin has already been packaged by a Critical-Care Air Transport Team (C-CATT); they've taken a look at his leg and decided that his body can endure the demands of flight. Using a preventative measure called a fasciotomy, Balad surgeons slit Heekin's leg wound open so that no air pockets will burst when pressure fluctuates in the airplane's cabin. Transport staffers drive Heekin and other injured soldiers to the runway where they are loaded in litters stacked three high.

Cradled in the metallic ribs of the cargo plane, up to thirty-six litter patients can be air-evaced. As with Balad, placement on the plane is ordered by severity, with the most severely wounded placed near the rear ramp exit. Intensive care patients are buried under hundreds of thousands of dollars' worth of state-of-the-art portable medical equipment. Barely any flesh is visible under the maze of tubing, monitors, and electrical cables. When I accompany a separate C-CATT mission out of Balad later in the week, I see an intensive care patient connected to nine different lines, four that deliver intravenous drugs, three that deliver sustenance, and two that remove waste. All of the other patients are placed with their feet facing the tail end of the plane, but because of the delicate nature of his brain injury, the ICU patient faces the opposite direction. In such a position, the angle of takeoff and descent is supposed to place less of a burden on brain pressure.

The idea of taking a fully functional, high-end intensive care unit and bolting it to the inside of an aircraft facilitates the military's medical care mantra: throughput. If you can continue to move the patient, you create ever-accessible medical care for patients who follow. The military is proud of its production-line styled medevac capabilities, which boast a long-standing history of devoted flight surgeons and cutting-edge research. Today, an active duty servicemember can make it from Balad to Germany and then America in the course of forty-eight hours; during Vietnam, the journey home took an average of twenty days. Most wounded Americans are already stateside before an Iraqi makes it out of Balad's intensive care units, for several reasons.

Iraqis don't wear body armor, so they come into Balad with bodies that look as though they've been thrown into a blender. The severity of their injuries requires, on average, twice as many surgeries as an American patient.[3] Iraqis arrive already malnourished, so their bodies take much longer to heal than well-fed Americans. In addition to a slow recovery period, doctors must consider the inadequacies Iraqis will face outside of Balad. When an American gets a feeding tube placed, for example, Balad surgeons choose a small tube for injectable nutritional shakes. Those supplements don't exist for Iraqis. Surgeons have to insert a tube several inches in circumference into Iraqi stomachs so that their families can grind up home-cooked meals and mash them into the wounded person's gut.

It is in this white, sterilized shipping container in the middle of a desert six thousand miles from my home that I see my first living human brain, Jassim's brain. I've seen brains pickled, bloodless, dried, and plastinated, but never aglow. For years, I've dodged other invitations to witness neurosurgery; I was content knowing the brain only through action and aberration. Now, with a child as the secret bearer, I can face the matter directly. I am ready to witness the brain ensouled.

In a back tent, I put on the surgeon's vestments—the scrub tops and bottoms, a borrowed pair of too-big rubber clogs—and I slip a baby-blue hairnet over my head. Outside the operating room, I run my arms under water, lather them in the immense basins, rinse them once more, and walk into the sparkling white room, the brightest in all of Iraq.

Melton is shaving Jassim's head delicately and evenly, as if he were a barber. He dips a wet washcloth against the scalp, soaking up blood and mud. Jassim's stubbled head reveals the damage. It has lacerations across the top, and a mangled flap of loose scalp dangles free. The top of his head must have bashed against something; perhaps he took a flying brick to his skull. At Jassim's leg, an orthopedic surgeon appraises the jutting black tibia. He glances at the bone, then at the tray of drill bits, and proceeds to set the leg in a sling.

The two surgeons work so intently that I hardly notice the child on the other table, a small girl about three years old. Right below the knee, her right leg has been cleanly amputated, the wound looking as perfectly round and damp as a tuna steak, with a protrusion of white bone in the middle. Just this morning a leg was there, and now it's gone. When she wakes up, her brain will tell her that she can feel her toes, and when she looks down to find out why they hurt so badly, her leg won't be there. Nobody will be able to explain where the lower half of her leg went, not in a way that will make sense to her.

I take a seat on a barstool directly behind Jassim's head and watch Melton debride the wound.

"You really gotta get things clean," he says. "The bacteria is in everything here, including the dirt. If it gets ahold of you, there's nothing we can do. It'll eat you up."

Melton runs a scalpel across and under the skin, then pulls a large piece of it off Jassim's head and lays it on the table near his ear. With the scalp cleared away, the underlying bone fracture is apparent. The glistening white skull looks cracked and cobbled at the very crown of the head, near the parietal lobe, at the point yogis call the *sahasrara* chakra. Those who attempt the ancient Tibetan practice of *pho-wa*, the art of dying, attempt to direct their consciousness out of the crowns of their heads during the moment of death. Practitioners claim to literally create a small opening in their skulls from the exercise. In their eyes, Jassim's injury must be an advantage. It should be easy for his spirit to depart. All too easy.

Melton uses a small set of pliers to pry the bone fragments away. At moments, he has to wrench the cracked bone free. A pile of small skull chips collects near the discarded scalp.

"Whatever you're doing, he doesn't like it," says the anesthesiologist suddenly. I look over at the monitor and notice Jassim's heart rate plummeting, down thirty beats per minute from a few moments ago.

"Give it to him little by little," Melton says without looking up. The anesthesiologist pushes on a syringe, and Jassim's heart pounds harder, faster.

When Melton's hand pulls away from Jassim's skull, I bow my head down and see circles of tissues within tissues: hair, skin, skull, marrow, dura. There, surrounded by brilliant white bone, is a glistening quarter-size ruby, Jassim's brain, slightly bruised, with droplets of blood dripping at the skull opening's edge. It is the shrine of Jassim's being, the temple where his thoughts take flight and dreams flash. It is Jassim, and not him, impossible, yet pulsing ever so faintly.

Jassim's brain is expected to swell, so Melton decides to drill a pressure monitor into his forehead. After the probe is set, Melton sews Jassim's scalp back on his head. The stitching pattern creates a thick scar with the unmistakable shape of a large letter W covering the crown of Jassim's head.

While comfortably bedded at Landstuhl Regional Medical Center, a comparatively luxurious Level IV trauma center in Germany, Sergeant Heekin received a First Response backpack that contained a number of carefully selected toiletries, fresh underwear, an expensive quilt, and a phone card. At around the same time, Jassim's father received a few rolls of gauze and a fifteen-minute lesson on how to dress his son's wounds and how to guard against infection. The bacteria is everywhere, they told him.

I leave Balad before I ever see Jassim regain consciousness, but I check in with Dr. Melton via e-mail with the hope that Jassim avoided palliative care.

"He is expected to make a good neurologic recovery," Melton writes. "He was conversant, eating, and drinking on discharge."

Melton's optimistic prognosis for Jassim's biological recovery, though, is deflated by the reality of Teff's input, which arrives in a different e-mail:

"Iraqi care for children is even more limited than it is for adults here," he writes. "The pediatric hospital at Medical City is either closed or offering limited services. Advanced pediatric intensive care is not available at any facility I am aware of. Though the American Level III facilities are doing their best to provide pediatric-certified providers and equipment, our resources are limited also. The children seldom have the stamina to survive infections and multitrauma insults like adults do."

Following a brief stay at Landstuhl, Sergeant Heekin was transported to Walter Reed Medical Center, where the military's pipeline of dreamcare has revealed nightmarish shortages replete throughout the entire VA and American healthcare system. While newspapers paraded photos of peeling wallpaper at army hospitals and portraits of disillusioned vets, Jassim began his own return home, a journey that didn't involve flying intensive care units, a war medal, rehabilitation planning, or brain injury case managers. There won't be a nonprofit to mail him a care package, or a foundation established for his cognitive therapy. If he can fend off the infections, Jassim might be able to relearn the route to school. When his friends ask him what happened to him, he will have to say something. God willing, his father will teach him never to utter the name of the hospital in the desert.

WOOD OF THE SUICIDES

for John Galusha

Any brain injury story can take a turn toward suicide.[1] Four months after my visit with Cheyenne, four months after he tells me about his snowboarding injury and his epiphany in the nursing home, he tries to kill himself. His mother calls and leaves a frantic message on my voice mail at work. The seizures were intensifying. Cheyenne had a knife, he had started drinking, he was upset about a girl. Now he's on a psychiatric ward at UCLA's medical center, where he might actually get some attention for his brain injury. Attempted suicide may be one of the best things that has ever happened to him; it's the kind of twisted truth that ends up making sense in the terrain of damaged brains. I call Cheyenne's home and say some words on his answering machine, but I don't know if he'll ever return the call. All I say is, "I want to hear about it."

Working with suicides exacts a toll. You have to consciously seek out light and life when you spend the bulk of your week in shadows. I've always counted on my friends to lead me out from the haze, and John was one of the most skilled life givers I knew. Over cocktails at a downtown bar, he once asked me how I liked working among the suicides. He was a painter and he had a curious mind and a misunderstood heart. I didn't want to tell him that I loved suicides, but I did. I could tell John anything and John listened. I told him that I loved suicides because they were honest.

"I can't blame someone for wanting to die," I told him. "It's a human impulse, like something inside of us suspects that death may be more bearable than life."

"You don't think it's a selfish, ugly thing, the way it tears families apart?" John asked me.

"It's horrific, one of the worst things a person can do," I told him. "But 'selfish' doesn't apply, even though that's the one word we like. It's the only thing we can say, because we're angry and we want something to explain it, something to take the blame."

I didn't know what John was asking, really.

Take the interstate and it's almost ten hours from Tulsa. Take the two-lane roads, mind the deer, and you can make it in seven. On your way up Highway 81 from Kansas, you'll pucker your face when the rank, steamy odor from the stockyards catches you downwind, but once you've crossed into Nebraska you'll hit pockets of cornfields so sweet that you'll swear it makes your hands stick to the steering wheel. You have to drive through six kinds of nowhere to arrive in

Norfolk, population 23,878. Daniel Harris lives in the seventh nowhere, Norfolk Regional Center, a mental health institution located a safe distance from the rest of town.

The contemporary name doesn't do much to hide the institution's history. Norfolk State Hospital for the Insane was built in 1888 and grew to house over a thousand patients in eighteen different buildings. Today, only Building 16 remains in use. Staffed by two psychiatrists, too many psychologists, and a handful of clinical workers, Building 16 houses sixty sex offenders, fifty-nine severe and persistent mentally ill patients, and one brain injury survivor, Daniel Harris. According to Nebraska's healthcare system, this is the only place in the state suited to treat Daniel, despite the fact that there isn't a single neurologist or neuropsychologist on staff.

Daniel is the one guy in the world whom I don't want to see, and I'm seeing him for the second time. I once evaluated a patient who removed her eyes with her own hands, I've driven around with a would-be presidential assassin, and I have shared small, enclosed rooms with cutters, gougers, pickers, and fire setters. Daniel is the only patient who frightens me. Although Daniel doesn't know it, he carries a ghost. Daniel is the same age as my friend John. Daniel had hopes of becoming a police officer, and John had breezed through the rank of Eagle Scout. Daniel went to college and left early. John went to college and left early. Daniel had a drinking problem; John made drinking his religion. They were both lousy with women. Not long after he turned thirty, Daniel hung himself, as did John. Daniel survived.

The sign marker for Norfolk Regional Center reads WORKING TOGETHER, GIVING OUR BEST, MAKING A POSITIVE DIFFERENCE. I drive past the sign, pull onto a dirt road, and then drive onto the campus using the service entrance. The road takes me past a dozen decrepit buildings, each of them a testament to the grand, overblown ideologies that birthed institutionalization. Most of the windows on the vacated buildings are knocked out or boarded up. I pass an atomic-age swingset that no longer holds swings, only chains, and then pass a maintenance building with a tall smokestack and a Ford pickup parked beside it. The driveway eventually leads me to the visitor parking spaces in front of Building 16. My car is the only one that occupies the empty spaces.

You have to fill out a visitor's request form before the old lady behind the security glass will let you pass through the metal detector and into the hospital. The first door, made from heavy, honey-colored wood, opens like all the patient doors, with an antique skeleton key. The lock makes a loud *clank* as it disengages, and a social worker opens the door and leads me back onto the floor. We immediately appear in the foyer by the nurse's station, a large box where patients line up to take their meds. Many of the patients are parked in various chairs, and some of them are shuffling down any of the three hallways. Daniel isn't among them. A majority of the patients are in sweat suits, while the techs and nurses wear street clothes. A patient with a frazzled gray shock of hair approaches me with her thin arms branched out and makes the sign for "busy" with her hands. I don't know the sign for "nervous," so I just nod at her and continue walking with the social worker.

We end up in an office at the end of the hall, and the social worker tells me he will get Daniel's records together for me. Daniel, he says, decided to take a bath today because he knew he would have a visitor. He shouldn't be much longer, the social worker tells me, and I tell him it's okay, that I'm in no rush. It's been two years since I last saw him, I say, I don't think ten minutes will make a difference. I'm not eager to see the effects of two years' worth of institutionalization on anybody, especially Daniel. I failed getting him help the first time, and he ended up here. I will probably fail this time as well. The social worker entertains me with a bit about the history of Norfolk, how Johnny Carson was born there and how he gave the town a bunch of money. He's in the middle of telling me about their high school football team when I turn around and Daniel walks in.

Shane Harris is a tall, lean roughneck—a compliment in Oklahoma. As a production foreman in the oil fields, Shane spends his week driving back and forth across the panhandle making sure that new wells produce and old ones keep pumping. He's been working the oil fields for nearly thirty years, and he looks it. All the skin left unprotected by his blue jeans and denim jacket is leathered red by years of sun. With his white hair, Shane Harris looks like a walking American flag and has the full, deep voice to match. I met up with him in his office north of Oklahoma City, a one-story industrial building right off the highway, in between large pastures.

After a relationship strained by addiction, Shane divorced Sarah Harris when Daniel was twelve, and although Daniel remained with Shane, Shane encouraged his son to keep in contact with his mother.

"Daniel and his mom always had a closeness," Shane says. "There was something about those two, you know?"

After the divorce, Sarah left Oklahoma City and moved to Omaha, Nebraska, to start a new life and a new cleaning business. Daniel would spend summers with her throughout his teenage years, but preferred the stability and freedom that life with his father afforded.

"We were close growing up. I felt sorry for him and let him get away with things he shouldn't have," Shane tells me. "That's where I have some guilt. Daniel began to rebel in a large way."

During his senior year, Daniel decided to move out of his father's house and live in Omaha with his mother. Once there, however, he reneged on his promise to complete high school. Daniel treated the time like a vacation and wandered from party to party, making friends and getting drunk. He showed up at Shane's door a year later, asking for his old room back. When Shane insisted he pay rent, it snapped Daniel's sense of responsibility back into place. Daniel joined the National Guard and turned his life around. Throughout his early twenties, Daniel earned his GED, found a job as a security officer, and eventually was accepted to the University of Nebraska at Omaha. He moved back in with his mother, and it was during his first attempt at higher education that clinical depression took hold of him. As soon as his freshman year ended, Daniel informed his father that he was coming home to visit, but the trip was cut short.

"When he called me, he had checked himself in down the road at a hospital in Oklahoma City," Shane says. "It's the first indication I got. At the time, I didn't have a lot of respect for psychiatrists. I told Daniel that if he had really wanted to commit suicide, he would have gone on and done it. I thought he was looking for attention and I didn't want to hear it. At the time, I didn't think it was that serious. He got out of the hospital for a few days and went straight back to Nebraska. That was as point blank as I've been with him. I don't know if that was a mistake or not."

. . .

I wrote an essay on suicide that won first prize in a local competition. It was about a job I once held as a psych tech, and I was asked to read it on a Sunday afternoon in April. April, of all months, John. I read the essay on suicide. I stood up behind the lectern, and I read out loud, and they gave me first prize, fifty dollars, and shook my hand. My daughter was in the audience, and though she was only nine then and had hardly paid attention to a word I said, she clapped also.

After the reading, I sat back down and listened to the other writers read their selections, and then Cherish and I stayed and ate cake and drank punch and made small talk with the attendees. I stayed there and chatted with a stranger about suicide and psychiatry and Ken Kesey and how I worked as a suicide sitter at times. I smiled when an old man, a retired doctor, approached me and congratulated me for how I handled suicide. I actually smiled, in my ignorance. I thought I had done something.

I drove home and it was evening already and the sky was grayish green, the color of leftover storms from the night before. I sat at my computer, the first chance all day, and I checked my e-mail, and the first one I saw was from my friend Sarah. It said: "Mike, call me." The e-mail below Sarah's was from Fran, a friend I rarely heard from. In the subject line was John's name. The e-mail said: "I don't know how to get ahold of you otherwise. Sorry to be the one to let you know, John hung himself sometime Saturday night. I don't know any details as of yet, but I will pass on any arrangements that I find out." Oh my god, I say, and Christy, my wife, asks what, and I say John is dead.

Christy knew John and loved him also, and she puts a hand to her mouth. She sits down to read my e-mails, and

I stand up and go into the backyard and I walk in circles. That's it. That's all I know to do. I wrote an essay on suicide that won first prize in a local competition, and I go home, and I face suicide, and all I can do is walk in circles. I make circles in my backyard until the green sky turns purple then gray then black all in what seems like seconds. It rains and John is a suicide and I keep making circles. I wrote an essay on suicide that won first prize in a local competition.

The encounter at the hospital created a rift between Daniel and Shane, and the two remained estranged until a tragedy forced them to reconnect more than a year later. As part of his coursework toward an undergraduate degree in criminal law, Daniel had opted to study the English court system in a study abroad program. He had already completed several weeks of study and was about to hop a train for a weekend sightseeing tour when a professor stopped him and delivered the news. Sarah Harris was dead.

Shane met his son at the airport in Omaha and helped him with all of his mother's arrangements. The coroner had determined that the cause of death was a fatal mixture of prescription drugs.

"I'm not going to say it was intentional," Shane tells me, "and I'm not going to say it wasn't. The truth is that we'll never really know."

At the time, Daniel had been living with the girl who would soon become his fiancée. Margaret was working toward a computer science degree, and she and Daniel had met through a mutual friend. Together Daniel, Shane, and Margaret collected Sarah's belongings and donated what they could. Daniel talked about quitting school and taking over his mother's business.

"I told him his mother wouldn't have wanted him to do that, but it took him a few days to make up his mind," Shane recalls. "Before I left, he told me that he was in love with Margaret but wasn't sure if he should marry her. We started talking more after that."

Daniel finished school and invited his father to his graduation, a day Shane Harris wasn't about to miss. He drove the eight hours to Omaha to attend Daniel's graduation and applauded and whistled when Daniel walked across the stage. As part of his graduation gift, Shane bought Daniel tickets to the College World Series, and the two passed innings sipping beer and talking about the future. Daniel complained about how difficult the job market was in Omaha. Margaret had already been offered a high-paying job, though, and it seemed like only a matter of time before he would find work also.

"I could tell there was already a little tension between the two of them when I visited," Shane says.

In the following year, Margaret bought a house in the Omaha suburbs while Daniel was forced to wait tables while looking for work. He increasingly turned to alcohol for consolation, but his drinking only increased the divide between him and Margaret. More than a year passed before Margaret decided that she couldn't deal with Daniel any longer. She had begged him to stop drinking and to find help, but he resisted. She asked him to move out of her house, and he complied. He found a friend's couch and told Margaret he would stop by in a week to get the remainder of his belongings.

On a Sunday morning in August, Daniel showed up to Margaret's house drunk and apathetic. Margaret led Daniel down to her basement, where she had collected his possessions. She left him alone while she went upstairs to continue with her cleaning. While she was gone, Daniel stepped up on a chair, knotted a rope around his neck, and fastened the other end to a crossbeam. He kicked the chair away, fell, and hung himself. The rope constricted around his neck, pinched closed the carotid artery, and stopped the blood flow into his brain. Daniel lost consciousness in seconds. Deprived of sustenance, his brain cells began withering away en masse. Just underneath his skull, pinpoint hemorrhages began to appear, sparking his brain's swelling response. As Daniel's brain tissue expanded, it began to smash the cerebellum and brain stem against the base of his skull, pinching off the already diminishing blood supply to other areas of his brain. As each moment passed, Daniel experienced losses in the thousands.

There are two routes a suicide takes into a hospital: the morgue or the emergency room. If the suicide survivor is conscious, the seriousness of the attempt is weighed in the emergency room. A fledgling social worker relies on various forms to screen for suicide risk, such as the Suicide Screening Checklist, but veteran intake nurses might take a much more intuitive and reliable approach: transference and countertransference. When we engage in casual conversation, we take part in an emotional tug-of-war. Sometimes we're trying to get a particular emotion or reaction; sometimes we willingly give an emotion or reaction when it's requested. During a standard hospital assessment, a potential suicide patient transfers a gamut of emotions to the assessor; the assessor then countertransfers with their own differing set of emotions.

When speaking to a depressed patient with relatively low suicide risk, a listener often countertransfers empathic, moist feelings. The patient wants empathy and wants someone to listen to them; they want life. But when faced with a truly serious suicide risk, countertransference takes on a markedly different tone. When speaking to a true suicide, the experience seems almost oppressively mundane, were it not for the topic at hand. Exchanges are often polite, but calculated in their distance, much like a business transaction. Words sound measured and disconnected, and even intimate information is relayed with an unnerving matter-of-factness. Pauses are heavy and long, and breaths are slow and mechanical.

Suicides are fierce in their silences. Life is so still, so quietly trapped inside them, they give the impression that they've already died and are just waiting around to make it official. Many nurses describe their encounters with suicides as vacuous, strained, or awkward—rarely is there an obvious hint of sorrow or gloom, as though the serious suicide has passed beyond the trifles of standard depressive conventions. When people are unaware of their countertransference, the instinctual response most have to a suicide is to leave the room immediately. The intake nurse, on the other hand, proceeds with admittance papers.

If the suicide is unconscious but alive, as Daniel Harris was after his attempt, the patient is stabilized and monitored until they succumb to death or become conscious. A suicide unwittingly floats through any number of branches in a hospital's carefully designed crisis plan: ICU, surgery, debriding tanks, cardiac care. At every stage of the process, they're still considered a suicide-in-progress. The threat of death remains as near as a paper clip, a shoestring, or a pillowcase. If a suicide emerges from unconsciousness with their memory intact, their response is usually a mixture of anger and embarrassment, and they may pull and rip at the tubes and cords sustaining their life. I've walked into a patient's room while she was attempting suicide with a bedsheet, and I felt as though I had walked in on her while she was using the bathroom. Her face was flushed with humiliation and rage. Suicide is an act more intimate than sex; no deliberate suicide wants to be caught courting death.

Daniel still had a pulse when the ambulance arrived; he had been hanging less than twenty minutes. At the University of Nebraska Medical Center, doctors placed a pressure monitor in Daniel's forehead to monitor swelling. A CAT scan revealed that no bone injuries were present, decreasing the likelihood of paralysis. By the time his father arrived at his bedside, Daniel was lying in an intensive care bed, unresponsive and requiring maximum care. He had a tracheotomy tube placed to support his breathing, he had a gastrostomy tube that allowed direct access to his stomach, and he had a Foley catheter for incontinence.

"It was very confusing because I was scared at first. This is gonna be hard for me to say," says Shane. His eyes are fixed in space, as if he's seeing Daniel in the bed all over again. "But when he was lying there, I wanted him to go ahead and die because I didn't want him to go through the pain. I didn't know what he was gonna go through, and if it wasn't any better, I didn't want him to live."

After John died, I attended his funeral, and I wrote a eulogy for him that ran in a local paper. In the following weeks, there was a rally of sentiment surrounding his death. Living Arts Gallery showed a retrospective of his works, and it was there that John's father thanked me for being a friend to John. I told him that John was beautiful and how could

I not be his friend. He cried openly and I looked at him and saw John and I wanted to cry also but couldn't. Instead I kept myself surrounded by John's art. I hung one painting in the living room, one in my bedroom, and set another on my desk. Each evening I fell asleep to the vision of his most fiery piece, and each morning I awoke to its swirl of dark reds and ambers.

A year later John's absence was still gnawing at me, so I decided to visit a Buddhist monastery in upstate New York. The monks must have known I stank of death; they avoided me most of the time while I sulked around the foggy grounds in silence. I had hoped that the solitude might provide a reprieve, but my thoughts circled John's suicide again and again. The maelstrom lasted days. Near the end of my visit, the abbess of the monastery called me into a private room and we sat on cushions facing each other. She had fair skin and kind eyes and wore a heavy black robe with an amber sash. The abbess looked at me with pure focus and attention. It felt peaceful just to have her eyes rest on me. She asked me if I had anything to say.

"A friend of mine killed himself a year ago," I told her. "And his death is in me now, more than his life."

"Was there a funeral?" she asked.

"Yes," I said. "But it wasn't the kind John would have tolerated."

"Did you hold your own ceremony for him?" she asked.

"No," I replied.

"I would like to do one with you," she said, "but I suspect there isn't time. This is something you must do for your friend."

"Why?" I asked. "Will this provide some sense of closure?"

"No," she said, "because what is it that opens and closes?"

She sat with me quietly until I stood up from the cushion and left the room.

Daniel's aunt was the first person from his family who called me. On the phone, she had explained to me that her nephew had an anoxic brain injury, and that he was in a nursing home in Nebraska, and that he was simply sitting in a room, watching TV and becoming restless. It pained her to see him so young and so abandoned. I told her to fax me the records. From his history and physical, I saw that he was thirty-one, left-handed, and of a usual state of health until his injury. I nearly got sick when I read the third line: *At the time, the patient was reportedly found hanging in a suicide attempt.* The medical records went into detail about how Daniel's throat was so swollen they couldn't even get the trach in at first, and how they required a 14-gauge needle to revitalize him.

He had spent one month at UNMC and was then transferred across the city to Immanuel Rehabilitation Center, where he received an intensive regime of physical, occupational, and speech therapy, which returned him to strong physical health over the course of a month. His cognition, however, was another problem. Early in his recovery, he was psychotic and claimed that his father and a visiting friend were vampires. He eventually learned to laugh at his delusions, but his memory proved to be seriously impaired. At the first demonstration of clear speech and thought, Daniel asked his father why he was in the hospital.

"You tried to kill yourself," Shane told him. "You were upset with Margaret and you hung yourself in her basement."

"Who is Margaret?" he asked.

Shane discovered that Daniel had retained only a few memories of Margaret from when they had first met.

"Why would I do that over her?" he asked.

Not only had Daniel lost most of his memories of Margaret, but he had no knowledge of ever having attended college, and he also did not know that his mother had died. The past six years were almost completely lost. On top of the gap in his long-term memory, Daniel couldn't retain new memories very well either. After the rehab hospital relocated him to a nursing home in North Bend, Shane took Daniel out on regular outings.

"In the course of one game of golf, he asked me the same three questions over sixty times," Shane says. "What did I do, why am I here, and where is my mom. He was like a little kid again."

At Shane's request, I reluctantly agreed to visit Daniel in North Bend. I knew that since Daniel had Nebraska Medicaid, I would have to tackle a bunch of policies in a state that didn't even have a functioning brain injury association. Daniel's biggest problem was that he was a brain-injured person who, although never violent, tended to get agitated easily, plus he would always be viewed as a suicide risk. There wasn't a single facility in Nebraska willing to treat both Daniel's brain injury and his behavioral issues. Daniel's case was a long shot, but if I managed to convince Nebraska to fund him at the place I worked, then other brain injury survivors might use Daniel's case as a precedent for realizing their own hopes of specialized rehabilitation. I couldn't say no to Daniel.

The nine-hour drive to North Bend was penance for not heeding the advice of the abbess; John was with me the entire time. I thought back to every meal we ever shared, every project we ever dreamed up, every argument we had, every joke we enjoyed, and every half-baked theory he ever proposed. John had an idea about fortune cookies: that if you tore the fortune in two, it always made for a better fortune. After plates of stir-fry, John handed me a fortune fragment that read "and go to the end of your thoughts," and I liked it so much, I tucked it into my wallet. A year and some months later, I sat in my car, trying to go to the end of my thoughts and finding John's death waiting. I thought about how his breath must have smelled of bourbon as he walked into his parents' garage, and how he tied a knot that he had learned in Boy Scouts, and how at four in the morning, in the middle of a raging thunderstorm, he set a ladder to a tree and ended his life.

John was already dying when I met him. In certain conversations, he alluded to the deliberate harshness of life as an artist. He memorized Yeats's poem "The Choice," the first few lines of which read:

The intellect of man is forced to choose
Perfection of the life, or of the work,
And if it take the second must refuse
A heavenly mansion, raging in the dark.

John insisted that a choice doesn't exist for the artist, that heavenly mansions offer little appeal over a healthy night rage. At the funeral, I learned that John had talked to the man who would be his coroner and had asked him questions about organ donation. It was the only time the coroner had met a customer before working on him.

The haunting drive had exhausted me. By the time I pulled up to the nursing home in North Bend, the sun was already setting, and I thought if I conducted my evaluation at night, Daniel might be tired and prefer to cut it short. I buzzed through the front door and was greeted by the night nurse, who led me down the bright fluorescent hallway to Daniel's room. She told me that Daniel had a light medication regime and for the most part never gave the staff any problems.

"He's just bored stiff here, the poor dear," she told me.

The nurse opened the door to his room, and I walked in and saw Daniel planted upright on his bed, wearing a maroon sweatshirt and blue jeans. His manner was a little rigid, but that could be just his personality. Despite spending the last few months in a hospital, Daniel looked healthy: thick, stylish hair, a wry, expressive face, discerning and attentive brown eyes. The only visual sign of his suicide attempt was the knotted pink tracheotomy scar in the crest of his throat. I kept my eyes on Daniel's face during the entire evaluation.

"Daniel, this is Mr. Mason from Brookhaven Hospital," the nurse told him. "He drove all the way up from Oklahoma to see you."

"I bet that sucked," Daniel said.

"You have no idea," I told him. I pulled a pen from my pocket and pulled out the first of my forms and began writing down my initial observations. "Why don't you tell me why you're here?"

"I was hoping you would tell me," he said. "All I know is that this place sucks and I want to go home."

"Where's home?" I asked.

"Mom's house," he said.

"Your mom died a while ago," I told him. Daniel didn't react, as I suspected he wouldn't. Some part of him guessed that his mom had died, even if he didn't actually remember the fact. I also suspected that he knew about his suicide.

"I just want to go home," he replied. "Why am I here?"

"Look, I don't really want to go into it," I said. "I'm trying to help you get home, and you can help me by answering some questions."

I opened the first of my forms and launched into a series of memory questions. When I asked him the date, Daniel guessed the month and day of the week correctly, but he was off by six years. As long as the questions involved something happening six years prior, he could answer easily. He listed off presidents; he talked about his family and the schools he attended. At one point, he told me that he felt strong as an ox and pulled back his sleeves and flexed a bicep at me. He would have made a good cop, I thought.

Daniel whizzed through the remainder of his cognitive testing, giving me the impression that he suffered severe impairments to his memory and moderate impairments to his judgment and insight. His facial expressions were dynamic and matched the content of his statements. Many of the other basics—his coordination, his conceptualization, his sequencing, his spatialization—revealed little to no impairments. In a casual conversation with Daniel at the movies or behind a desk, you might never pick up on the fact that he had a serious brain injury. I knew right away that if Nebraska did agree to pay for his rehab, the intensive in-patient portion of the rehabilitation would last only a matter of weeks before he would step down into a group home, where he would reside no more than two or three months before he would be ready to return home. He was a lucky one, I thought. Damned lucky.

Daniel lost all memory of the events that drove him to suicide. Instead of facing a future of serious impairment, he might achieve a higher level of independence than most of the patients I meet. He would probably be able to hold down a basic job within a year's time, and depending on how aggressively he took his rehab, he might be able to establish new relationships and achieve new hopes and dreams. Daniel was going to make it through this, I thought. Before I left, he told me that he bet he could find a job as a security guard someday, and I told him that it might be a good job for him. I shook his hand, walked out of the nursing home, and exhaled.

Daniel's recovery depended on a number of different factors, but perhaps the most significant one in his case was a matter of geography. Get a serious head injury in Council Bluffs, Iowa, a town that borders Omaha, and odds are that you'll qualify for a special healthcare waiver that will allow you to access a number of brain injury services and programs. Depending on the severity of your case, the state of Iowa may pay for years of rehabilitation, without any significant cost to you or your family. Cross the street into downtown Omaha, get the wrong kind of brain injury, and you're fucked six ways from Sunday—a predicament that isn't exclusive to Nebraska. Most case managers who deal with brain injury know that dozens of states don't have laws, policies, or regulations that come close to acknowledging, much less addressing, the needs of brain injury survivors.

Nebraska has one long-term rehab that primarily treats individuals with brain injuries, Quality Living, Inc. It's an enormous facility, the size of a college campus, with all the amenities a disabled person could hope for: an indoor heated pool, a state-of-the-art technology center, a thriving vocational program, and private apartments. The big problem is that its perks also put it in nationwide demand, forcing the place to operate at capacity. Fewer than half of its beds are occupied by Nebraskans. Another complication is that the facility isn't currently equipped to handle any of the severe behavioral problems that result from brain injury, which excludes a large number of survivors.

The state of Nebraska asked QLI to consider accepting Daniel Harris into their brain injury rehabilitation program. QLI rejected Daniel's request for admission, stating, "Daniel's need for structure is stronger than what our open campus is able to provide at this time." QLI recognized that Daniel was at a high suicide risk, that he was prone to agitation, and that their facility wasn't the right place. They communicated this to Nebraska the very first time they evaluated Daniel.

A few weeks after my evaluation at the nursing home, Daniel's patience wore thin, and he eventually acted out. In a fit of frustration, he pushed another resident in a wheelchair out of his way, and then tore up his college diploma. While Daniel did not harm the patient, the outburst was enough to get him transferred to a psychiatric ward in Omaha until another placement could be found. Before and during this time, I requested that Nebraska create an exception to policy that would allow Daniel to receive treatment at Brookhaven—a common protocol by which some other states agree to send us their brain injury survivors. In a maddening display of absurdity, my request was summarily denied. As far as Nebraska was concerned, there was only one place in the entire country that would house Daniel: Norfolk Regional Center.

Shane Harris visited Daniel at Norfolk Regional Center shortly after his admission and expressed worry that his son wasn't in the right place. His concerns were only validated when both a social worker and a psychiatrist at Norfolk provided him with the following letter printed and signed on State of Nebraska letterhead:

> Daniel Harris was admitted to Norfolk Regional Center as a transfer from a psychiatric unit in Omaha. He carries a diagnosis of 294.0 Amnesic Disorder due to Brain Anoxia. Shane Harris serves as his son's legal guardian.
>
> At this time, it is believed that Mr. Harris could benefit from more specific treatment regarding his neurological impairment. This rehabilitation process could be accomplished through a transfer to Brookhaven Hospital in Tulsa, Oklahoma. The family is very concerned and interested in this patient's treatment and rehabilitation. We feel this transfer would be in the best interest of the patient and his family.

Two months later, Nebraska responded to Norfolk's letter with a three-page fax citing a list of policies that Daniel's transfer would violate. The letter ended with the following empty hope:

> Quality Living, Inc. of Omaha, NE staff will continue to monitor Mr. Harris's situation and when he maintains non-aggressive and non-agitated behaviors on an open ward for a period of time, will reconsider their denial of admission. If QLI staff determines that an admission is appropriate, they will notify our staff for payment authorization. We will reconsider based on the 471 NAC regulations previously cited.

Since the issuance of that document, QLI has evaluated Daniel Harris on several occasions and repeatedly denied his transfer for the same reason cited in their initial decline. Daniel doesn't fit there, or anywhere else in Nebraska.

Modern psychiatry overinvests its resources into researching the physiological factors surrounding suicide, but leaves suicidal theory and speculation in the hands of psychologists and psychotherapists who have done little to alter the course of suicide in contemporary life. One of the richest sources for a deeper understanding of suicide comes not from any contemporary journal or research, but from the weighted stanzas of a medieval masterpiece, Dante's *Divine Comedy*. Dante's version of hell as a series of rings within rings creates a structurally solid and universal portrait of human desire and suffering. The same souls that populated his hell back then crowd our streets today. In Canto XIII of *Inferno*, Virgil leads Dante to the second ring of the seventh circle, the Wood of the Suicides. He describes the place as a nightmarish forest where the souls of suicides spend eternity.

"On every side," Dante wrote, "I hear wailing voices grieve."

At first, Dante thinks that the suicides are crying out from behind the trees, but the moans well from the trees themselves. When Virgil encourages Dante to break a limb from a tree, Dante recoils at the flood of blood and words that follow. The tree begins to cry out, recounting the horrors of its previous life until finally the flow of blood-sap thickens and stops, silencing the story in midtale. Dante's suicides find the freedom to express their anguish only after they've been torn and broken, and not a moment earlier.

The stanzas in Wood of the Suicides reveal such sensitivity and insight that it's easy to conclude that Dante knew suicide from the inside out. His muse, Beatrice, died at the age of twenty-four, leaving Dante in the throes of a suicidal sorrow. Scholars suggest that the thought of abandoning his children prevented Dante from killing himself, but anyone who works with suicides sees the academia at work behind that, or any other, line of reasoning. Some suicides are avoided, some are not, and none of us know why.

Throughout the year I placed intermittent calls to various Nebraska healthcare professionals. At one point, I sat at a roundtable staffed by a dozen central-state head injury administrators and directors, three of them Nebraskans, and brought up Daniel's case. Nobody could figure a way out for Daniel—though several of them joked he should move to Iowa. Nebraska's policies would not allow Daniel access to the services he needed, and the state had worn out Shane Harris with their consistent rejections. During my travels I was able to catch up with Shane at his office, and I asked him what he had learned from dealing with Nebraska over the past two years.

"I've been to Nebraska probably twenty times in the last two years," he began, "but in the last six months, I've only been there once. I'm at my wit's end. I feel like writing Nebraska a letter saying, 'Take care of it, I don't want to deal with it anymore.' But Daniel is my son, I can't do that.

"Nebraska is Daniel's home," he said. "He's been there all this time and paid his taxes. They are responsible for him, but when it came to the point of bringing him to his family in Oklahoma, their answer is just no. You make phone calls, you fill out papers, you sign releases, and it all seems like it's for nothing. They always got rules and regulations that they state. I feel powerless against Nebraska. I've been told that if I can get Daniel a lawyer, and start this whole process again, then there might be a chance of getting him out of there. But the same person who told me that says there are other people in Nebraska trying to do the same thing. I feel like I don't want to waste my time doing it."

Daniel's brain injury forced Shane to devote an inordinate amount of time challenging Nebraska's health and human services system. He likened it to taking on a full-time job on top of the one he already had, and he'd finally reached his threshold.

"I've got a big box full of papers on Daniel and I just walk past it now," said Shane. "I'm not proud of that. My mom is dying, that's a priority. My daughter is living with me and she's working hard to get through school, and I've got a grandson. Daniel is where Daniel is because he did what he did. This is something he will never quit paying the consequences for, and now he has me paying them, too. Yeah, I get angry. Maybe I didn't raise him the best, I don't know. I tried. I'm not going to say I am the best father in the world. I don't know. But I do know that there's a little five-year-old kid in my house, he's brand-new, and he can probably do better than me. But Daniel can't."

I asked Shane if he thought I might be able to contact anyone else at the state level for him, but he shook his head.

"They say they want to help, but honestly, they don't want to help," Shane said. "They'll tell you why they can't help. It doesn't matter what you throw at them, they turn the page over and there's another regulation. There's another piece of paper you have to fill out before it goes back to the same person who said 'no' to begin with. They told me that if I really wanted him home, I should just go over there and pick him up. And if I were a millionaire, sure I would do that. But I can't afford to quit work and stay at home with him by myself. I can't afford to pay someone to watch him all the time. So how the hell can they expect me to just take him home? Daniel could walk out a door, down that street, and never make it back because he just doesn't know. I can't take care of that. I wish I could, but I can't."

Shane's frustrations tunneled inside me as he spoke. I watched him shake his head in confusion and fidget with the pen on his desk. He needed a break, and the only way he knew how to get one was to stop fighting for Daniel. I knew that my schedule would require me to visit Nebraska soon, and that I would be within a couple of hours' driving distance of Norfolk Regional Center. Although I knew it wouldn't make much of a difference to Daniel, I asked Shane if he wouldn't mind me visiting his son.

"I might see something I haven't seen before," I told him. He thanked me for my willingness, and I nodded back, regretting the offer the moment I spoke it.

Before leaving his office, Shane gave me a last word of caution about visiting Norfolk.

"You screw your life up, and it's not pretty," he said. "It's hell. It's hell on earth."

It has been two years since I laid eyes on Daniel. I hardly recognize the man before me. Daniel's face is notched with acne scars, his skin is dry and pale as balsa, his hair is a bushy mess, and he has dried spittle in the corners of his mouth. He has gained about thirty pounds and moves at a glacial pace toward the chair near me.

"Daniel," says the social worker, "Mr. Mason is here to visit you. He came all the way from Oklahoma."

"Oh," Daniel says as he takes the seat.

I hold out my hand to Daniel and notice his fingers trembling as he slowly stretches his arm toward me. He has no grip. Daniel hunches over in his chair and stares blankly at the table in front of him.

"Daniel," I begin, "I know you don't remember me, but I visited you a couple of years ago. I just dropped by to see how you're doing."

"That sucks," he says.

"Everything *sucks*," the social worker comments. "That's Daniel's favorite word now."

"Could you tell me how long you've been here?" I say.

"A month," he says. It is in fact his sixteenth month at Norfolk.

"Could you tell me where you are?"

"I don't know," he says, drawing each syllable out slowly.

"What about the state? Do you know what state this is?" I ask.

"I don't know," he says. "I think Nebraska."

"What's it like here?" I ask him.

"Everything seems weird," he says. "It just doesn't feel right. It feels dreamy. Things happen and I don't know why they happen."

"So what do you do here?"

"How would I know?" he says. "It's my first day here."

Daniel's tone is flat and despondent; I ache just to hear him talk. The social worker hands him a bag of popcorn as we speak, and instead of putting his hand into the bag, Daniel cranes his neck down to the bag and pecks at the popcorn with his lips. His motor coordination has deteriorated significantly; he won't be able to hit a golf ball, much less walk an entire game of golf anymore. The remainder of our conversation consists mainly of monosyllabic replies from Daniel, except when I ask him about his plans for the future.

"I want to go to Oklahoma," he says.

"Do you want to go back to work?"

"No," he says. "I want to sit around."

I ask the social worker for information about Daniel's behavior while he has been living at Norfolk, and while he tells me about Daniel's history I thumb through his chart, curious at what medications they have him on. The med sheet reads like a list you'd see on a person suffering extreme mental illness: clonazepam, divalproex, duloxetine, memantine, risperidone, and lorazepam, to name a few. The only serious diagnosis on Daniel's chart is that of amnesic disorder, which won't even meet the criteria for admission in shorter-term psychiatric hospitals. If I were to take just one morning cycle of Daniel's medication, it's likely I wouldn't wake up for days, possibly ever.

Part of the physician's orders are a list of activities granting Daniel small liberties that he could not otherwise enjoy without a doctor's approval. Daniel is allowed to leave his room door open, he is allowed to visit the courtyard, he can make two phone calls per week, and he is allowed to shower three times a week, with the proper supervision. The progress notes are filled with telling encounters that reveal the effects of institutionalization. Daniel regularly asks for permission to call his mother. He sleeps in all morning long. He turns down meals almost daily. He isolates as much as possible. The social worker tells me that while Daniel isn't considered a behavioral problem, he has lashed out by yelling at others and hitting them. I don't know that I would act any more civil if the same thing happened to me.

"Tell me." I turn to the social worker. "Are there any plans to change Daniel's program here?"

"No," he says, and looks down. "There's no plan to do anything different."

The heaviness of Building 16 is taking root in me, so I say goodbye to Daniel, knowing that by this time tomorrow, he will not even know I exist. The social worker escorts me through the back hallways and out into the gray sky, and I climb into my car. Before starting it, I check my messages and see that Cheyenne has called. His message says that he is okay, that he feels a little stupid and embarrassed by the ordeal over his suicide attempt, and that he is hoping he can talk to me about it. I sit there for a few moments, wondering whether I should call him back right away or not.

Instead of starting the car, I swing open the door and leave my belongings inside. I start my way up a long sidewalk that circles the entire institution. The Norfolk grounds are quiet and desolate, and as I pass each building, the numbers ascend. I stroll my way past the old swingset, and I feel my chest sink as I approach the old stone building that first served as Norfolk State Hospital for the Insane, Building 1. I peek inside one of the broken windows and tense my face against the souring stench of mildew and urine. Paint peels from the walls and ceiling, broken floor tiles and bottles litter the hallways, and boxes filled with papers and office detritus line the corridor. The stories of the thousands of mentally ill and disabled once abandoned to Building 1 now lie entombed in cardboard, abandoned once more.

As the buildings vanish behind me, I mull over different scenarios that might get Daniel out of Norfolk. There is a strong chance that waning funding for the hospital may force it to close, but then Daniel will just be transferred to the other state hospital. I could start an aggressive hunt for a civil attorney who might take up Daniel's cause pro bono, but I doubt that Shane Harris has the energy or the time to start up a new appeal at this point. I see no end to Daniel's limbo, no end that doesn't involve death.

As I circle my way back to the car, cutting back across a great lawn, my thoughts return to John, and how it seems incredibly disrespectful for me to weigh his death against Daniel's survival. So far I've managed to avoid holding a ceremony for John, fearing that it would mean the end of him, as though John were something that simply started and ended. There, on a simple bench, I decide to hold my ceremony. I pull out a small scrap of paper I have kept tucked in my wallet, one of the few reminders I have left, the half-fortune that he gave me. I crumple the paper into a small wad, lower to my knees, and push the paper as far as I can into the soft ground. When the fortune disappears, I smooth the spot over with my palm and whisper the lines from an evening prayer I heard a lone monk chant in the still hall at the monastery:

Let me respectfully remind you,
Life and death are of supreme importance.
Time passes by and opportunities are lost.
Each of us should strive to awaken.
Take heed:
Do not squander your life.

I touch the grass once more and I raise myself up.

CONCLUSION

The September sun graced me with golden light the entire drive up from Dallas, and I kept my windows down to let in the last remnants of summer. Pulling into the driveway, I looked a little windblown and happy already, so I was all the more receptive to my wife's suspicious smile. Although my arms were full of travel gear, Christy grabbed everything out of my hands and set it all down on the front porch. She threw her arms around my neck, leaned into me, and looked into my eyes. She said that the test showed two lines, undeniably solid. I smiled and spun her around and kissed her and we stayed there, circling each other slowly, for minutes, without saying a word.

She guessed that the baby would come late May or early June, and we decided we wouldn't tell anyone, not until after the first trimester. Last year, at around the same time, Christy had miscarried just weeks after we broke the news, and the disappointment undid our daughter, Cherish. At just ten years old, she had a tough time understanding how a baby could simply disappear, without so much as a rise of belly or a whisper of movement. If this pregnancy went differently, Christy told me, she wanted to break the news around Thanksgiving.

In early November, though, Christy began spotting. I was in a miscellaneous hotel room an entire nighttime away, and I sat quietly on my bed and listened to her weep through the receiver. I told her to call her father, a physician, and let him in on our secret. When I returned home, she told me that she had followed her dad's advice and had already seen her obstetrician, and that there was still a rapid little heartbeat. That night, I began sleeping with my hand just below her navel, cupping the life we made.

A complication called polyhydramnios affects 2 percent of pregnant women. The body produces too much amniotic fluid, placing strain on the placenta and surrounding tissue. The placenta can rupture, and depending on the stage of the pregnancy, injury and death can result. When the doctor informed Christy of the condition, I lowered my head. Even though she had already begun her third trimester, she still ran serious risks due to the rapid buildup of pressure.

The doctor explained that the excess fluid wouldn't harm the baby directly, but it often causes the umbilical cord to bunch around the bottom of the uterus. If the water were to break with the baby in head-down position, the baby's body would clamp the umbilical cord closed, cutting off precious blood supply. In such an event, the baby needs to be delivered within five minutes; any delay could initiate the horrific progress of an anoxic brain injury. Doctors often recommend a Cesarean birth to offset the risks of polyhydramnios, but ultimately, the choice remains with the mother.

Christy had hoped for a natural birth, and she pressed the doctor for more information. She wanted to know if she was in immediate danger, and he said no. She asked him if a vaginal birth was automatically ruled out, and he said let's wait. While she talked with the doctor, she didn't notice the color draining from my cheeks. I've met these children, the ones who survived brain damage during birth. They arrive into the world comatose or vegetative; they are engaged in a vicious fight for their lives from the first moment. Of all the types of brain injury, birth-related brain damage struck me as the most unjust.

There, in the quiet hold of the womb, my child sat in the crosshairs of the most catastrophic injury. A small tear in a thin membrane could issue massive devastation or death. Before that day, brain injury had always been business; now it crossed into the personal. I felt violated by the intrusion, as though working with survivors had somehow inoculated my family and me from the risk of injury. With the development of the complication, the line had been erased.

On a Friday in June two weeks before the due date, Christy and I went in for another consultation with the doctor. He did an ultrasound and scratched a few measurements into his chart. The polyhydramnios had advanced into a critical phase, he told us.

"I'm not telling you this to alarm you," the doctor said, "but you should know that in a very few cases, the fetus gets the cord wrapped around its neck. If that happens and your water breaks, the baby will probably die. There's just no way to get to it in time. A best-case scenario would be severe brain injury."

I asked him the odds, and he said that in his experience, it's one in a thousand. I turned to my wife and I told her that I felt we simply could not take the risk. She agreed, and the doctor immediately weighed in.

"Great," he said, sounding relieved. "Let's do it today then. I'll see you over at the hospital in two hours."

Christy and I looked at each other wide-eyed and left the doctor's office in a frenzy. In the car, we called our families. My mother squealed in delight and her mother sparked a long succession of phone calls. At home, we gathered her belongings, and in the middle of packing, Christy froze.

"What?" I asked.

"Nothing," she said, "but let's leave now."

Her contractions had started. With her labor underway, each minute held the potential for her water to break. We lived within six minutes of the hospital; I made it safely in four.

I'm standing by my wife's head, disinterested in the gore unleashed by the flitting scalpels. Christy can't see and can't feel anything that's happening, so I report back to her that her lap is filling up with water, and now there's a little wading pool between her knees. I'm waiting for the proverbial first scream, but instead I hear the doctor's voice.

"Looks like the cord was wrapped around her neck," he says, looking down. I see his wrists make an unwinding motion, then he holds up a little body, blue and still as stone, with little arms and legs curled skyward. The floor drops out from under me. Gone is the white operating room. Everyone in the waiting room outside vanishes. I no longer hear my wife's voice. I am suspended, alone with the image of my daughter, and I am the most helpless person in the world. I thought a Cesarean would prevent this from happening.

It turns out I was right.

Just as casually, the doctor taps her feet. The baby's mouth pulls open, the legs crank back and kick, her little hands clench in a tiny rage. Her cry pierces the room, a strong, fierce little sound, and in rushes a pink bloom through her waxy flesh. She is alive and the universe is glorious and bursting and it's all happening to me, to us, to the entire world. The room snaps into place around me; the walls and the doctors and the entire hospital trumpet with light and I hear my wife's voice.

"Michael?"

"Christy," I say, "she's so beautiful!"

At the end of the case manager's arc of responsibility is discharge planning. We're supposed to know where a patient will go once they've completed treatment, and a new case manager is designated to pick up duties. They will become the patient's new voice and eyes, their representative in the human services or healthcare system. At least that's how it's supposed to happen. In reality, patients often return to environments that are ill-equipped to accommodate their needs. In the hospital I work at, the bulk of discharge planning falls on the shoulders of our utilization review director, and so I'm often left only tangentially aware of what becomes of our patients. Sometimes they call, just to check in, and other times their trail is too hard to track. I like to think they have somehow found their own way, and that they don't need the hospital's help anymore.

The survivors leave, and when they are gone, I place their files into a large drawer and lock the cabinet. Every so often a patient reemerges in my caseload and I pull open the drawer. When I see the file again, the patient's history reads like an incomplete chapter in an epic saga. So much life transpires. With several hundred cases under my belt, and several hundred waiting, it's difficult for me to keep names and details matched. It's also a relief to let go of the faces, to free myself from further involvement.

Although I may lose track of the case, each one contributes to an overall effect on me. Everything I once thought of as fixed and immutable about myself has now become suspect. The palace of memory can burn to the ground, and Julie Meyer still stands with her arms around her notebook, smiling and smoking a cigarette in the sunlight. With just enough pressure, time and space relinquish their grip, but it doesn't stop Pony Soldier or Doug Bearden from taking family vacations. Innocence can beget violence and entire educations must be reclaimed, but those hurdles haven't slowed the Larson family in the slightest—or, as in Asya Schween's case, they might even propel you into success. Daniel Harris's injury holds him in limbo, but his own impairments may end up helping him outlast the system that confines him. Carrying your own body can become your greatest burden, but to Rob Rabe, it's just healthy exercise. Charm and agility can melt into insecurity and desperation, but a change of temperament hasn't stopped Cheyenne Emerick from falling in love with a tall, gorgeous woman who seems to accept him no matter what. Their brains have undergone irrevocable change, but their humanity abides.

. . .

Tomorrow morning, I hop a 7:00 a.m. flight to Albuquerque. I'm going to see a young man injured in an industrial accident. He can't feed himself and he wobbles into the walls when he tries to walk, but a week from now, his social worker tells me, he's supposed to leave the hospital. There isn't a bed within five hundred miles for him. I'm supposed to meet with his family, and when I show up, I'm going to disappoint them. I will tell them the real scenario, the one that nobody else has explained to them, and I will sit and watch as their smiles collapse and their eyes lift to the ceiling and turn red.

A few hours later, I will catch a late flight back home and consider the short trip a mercy. I will walk through my front door, set my satchel down in the back room near my desk, and kiss my two sleeping girls. As I undress, I will catch a whiff of the stale hospital odor stuck in my clothes, and I will wonder what he is doing, what it feels like to be waiting in between the world you left and the world you want. When I turn off the light and climb into bed, my wife won't ask me a word about the man I saw; she will simply curl up behind me and hook her arm around my chest, and I will gaze into the dark.

APPENDIX: MY BREAKFAST WITH MARILYN

She lives in the Boston suburbs, has grown children, works in a hospital, and has all the charisma and energy of a woman half her age. In high heels and a smart business suit, Marilyn Price Spivack makes an unlikely rock star, but in brain injury circles, that's exactly the image she conjures. In 1975 Marilyn's then-teenage daughter Debby sustained a brain injury in an automobile accident, and the event immediately plunged Marilyn into a life of activism.

"I like to say head injury happens to every person in the family," Marilyn says. "There isn't anyone in the family whose role isn't changed. It does not enhance lifestyle."

Outraged by the subsequent lack of understanding, support, and services for Debby, Marilyn established the National Head Injury Foundation, which later became the

Brain Injury Association of America. Millions of brain injury survivors worldwide owe their lives to her work.

"My husband told me I had a tiger by the tail." She smiles, recollecting the earlier days of the organization. "Boy, was he right!"

During our breakfast overlooking Boston Harbor, Marilyn and I talk about the current condition of brain injury in America. It is an election day, and so our conversation is charged with guesses about how our government might respond to the numerous hardships created by brain injury. Marilyn has just finished collaborating on a brain injury–specific report by the Institute of Medicine that reads like an indictment of the American health and human services system.

The quality and coordination of post-acute TBI services systems remains inadequate, although progress has been made in some states, the report states. *Many people with TBI experience persistent, lifelong disabilities. For these individuals, and their caregivers, finding needed services is, far too often, an overwhelming logistical, financial, and psychological challenge.*

"Brain injury looks very little different today than it did yesterday," Marilyn tells me. "Outcomes can be very different, but today the scene is weaker than it was in 1975 because of our public policy and healthcare structure. I do believe that in the past thirty years we have made some very significant progress in the areas of neuropharmacology and enhanced diagnostics and getting people care sooner, but long-term opportunity is much less hopeful today than it was when I began."

The Iraq War and the subsequent treatment of brain-injured troops have only served to underscore Marilyn's insights. The military has put a tremendous amount of funding and research into the initial phases of brain injury care.

They are saving lives at an extraordinary rate, but the success comes at a significant cost. Because of the increasing severity of the injuries, estimates place the lifetime costs for a brain-injured veteran at over ten million dollars.

Without a strong infrastructure in place to care for its own civilians, the U.S. health system is ill prepared to meet the needs of wounded troops. The Veterans Affairs system, which has been shrinking over the past few decades, is scrambling to meet the demands of brain-injured soldiers. The Iraq War has shed light on the hardships that all brain injury survivors have lived with for decades. With well over a million new civilian brain injuries occurring each year, America can no longer afford to ignore the problem.

"Congress has not increased its appropriations for rehabilitation research and services in years," Marilyn says. "I think we have made a commitment in this country, and in all Western civilization, that we will spend whatever it takes in pharmacology and technology to save people. Prevent fatality, prevent mortality. They speak to caring, but meanwhile services are being cut and access is being cut. When we first began this effort, everybody talked about the quality of life and maximizing functionality. It costs money. And a lifetime of commitment."

While Marilyn speaks, I can't help but think of the hundreds of cases I've worked on. Not a single one hasn't dealt with some major social obstacle. Even with the exceptionally rare cases where funding wasn't a primary issue, there are still tremendous hardships caused by communities that simply don't have resources to utilize. How do you get neuropsychological testing in Dodge City, Kansas? Who will spend a couple of months to reteach you the bus routes in Detroit or Las Vegas? The problems are not unique to America. What service can you garner in Cairo, Mumbai, Helsinki,

Torino, or Bogotá that will help you purchase the right medications, find a job suited to your current cognitive level, or simply assure your safety?

Although the Institute of Medicine report provided a clear statement of recommended changes to American health and human service systems, Marilyn tells me that it received virtually no response from government officials. The Brain Injury Association of America continues to operate with a pittance compared to organizations that serve similar-size demographics. The plight of our brain-injured citizens demands that we finally address their concerns in a definitive and encompassing manner. The response will require nothing less than a thorough evaluation of American healthcare and the implementation of systems that can absorb the monumental costs associated with brain injury. Communities must be reinvented around the needs of the disabled, and courts must police hostile insurance practices.

Without a doctor's degree or any official license, Marilyn Price Spivack has accomplished more for brain injury survivors than a thousand surgeons and lawmakers. She is living proof that you can grab tigers by the tail and emerge victorious. For thousands of brain injury professionals, Marilyn is the face of brain injury in America, but for Marilyn herself, that face belongs to her daughter Debby, who remains disabled yet insuppressibly high-spirited.

After our breakfast, Marilyn takes me up to Spaulding Rehabilitation Hospital, where Debby is currently receiving care. Debby welcomes me into her room with a magical, wide-armed hello. More than thirty years after her injury, Debby continues to make gains in her life, all because of the focus of one person, her mother. I'm happy to report that Debby's face, the face that changed brain injury in America, is as beautiful as any I have ever seen.

NOTES AND SOURCES

THE HERMIT OF HOLLYWOOD BOULEVARD

1. Fabricius Guilhelmus Hildanus, *Opera observationum et curationum medico-chirurgicarum*. Frankfurt, 1646. In Owsei Temkin, *The Falling Sickness* (Baltimore: The Johns Hopkins Press, 1945).
2. William James, *The Varieties of Religious Experience* (New York: Longman, Green & Co., 1902).
3. Psychogenic nonepileptic seizures (PNES), also known as pseudo-seizures, are psychological in origin and given a stamp of validation in the *DSM-IV.* Most PNES are categorized as a conversion disorder, a type of disorder falling under somatoform disorder. In plain English, this means that the person having the PNES is not conscious of the fact that they're faking the seizure. If they do know that they're faking the PNES, then the diagnosis falls under either a factitious or malingering disorder. PNES comprise about 20 to 30 percent of referrals in epilepsy clinics, with about 50 to 70 percent of patients becoming seizure-free once their bluff is called.

4. The website epilepsyfoundation.org has a well-organized protocol for aiding a person suffering a seizure. It's important to remember that a person cannot do anything to shorten or stop a seizure. If possible, use a jacket or other item of clothing under the person's head to minimize injury. Don't try to force the mouth open, and don't attempt CPR unless you're certain that breathing has stopped. Do turn a person on their side, as it may help clear their airway.

A PRISONER OF THE PRESENT

1. In 1998, engineers for the National Highway Traffic Safety Administration (NHTSA) slammed a three-thousand-pound concrete barrier into the side of a brand-new Mercury Tracer at a speed of thirty-eight miles per hour. The safety engineers measured the remaining survivability space, gauged the forces reported by the crash test dummies, tabulated their data, and assigned their ratings on a one-to-five scale, five stars being the safest. The Mercury Tracer earned a three-star front and side rating, indicating that in a similar accident, an individual stands a 20 to 35 percent chance of serious injury.

    Hybrid III, the dummy driver of the Tracer, received a Head Injury Criterion (HIC) rating of 681, just a few points shy of "marginal." Calculated with a complex set of algorithms, the HIC rating exists as an addendum to the NHTSA rating. The NHTSA rating tells you what your odds are for experiencing a severe injury, and the HIC rating tells you the odds of having a severe head injury regardless of what happened to the rest of your body. Monroney stickers, which are required on every new car, display information about the vehicle's make, model, price, place of assembly, fuel efficiency, and optional equipment. Until 2006, only Honda voluntarily included safety rating information on their Monroney stickers. As of November 2006, all automobiles in America must include a five-star government safety rating (if available) on their Monroney stickers. The most readily available source for auto safety ratings can be found at www.safercar.gov.
2. The low-hanging "fruit" Barry mentioned made me curious. I decided to put a call in to Dr. Robert Hubbard, a professor of biomechanical engineering at Michigan State University. Following the racing-related death of a friend in the eighties, Hubbard and

his brother-in-law invented the Head and Neck Support (HANS) device, a restraint system now compulsory in most professional racing events.

The HANS device looks like a neck brace that belongs on a hospitalized patient, not on a stock car driver. After the driver has donned a full-body fireproof suit, he climbs into the side window and into a carbon-fiber seat that surrounds the torso and keeps legs from flailing. While seated, the driver then straps on the yokelike HANS, the arms of which rest on each shoulder and join at a collar behind the driver's helmet. Two tethers extend from the collar and attach to each side of the driver's helmet. The seat belt, a five-, six-, or seven-point restraint, is an over-the-head design that straddles each HANS arm and buckles together at the driver's sternum. Properly strapped into the cockpit, the driver can't budge an inch of his torso or pelvis. Only his arms and legs have an appreciable range of motion; he can barely turn his head side to side about forty-five degrees. Confined, the driver is now ready to take a thousand left turns. He must rely on radio contact from several spotters planted in the stands to guide him through the crowded cluster of cars as they collectively speed down the track at speeds approaching two hundred miles per hour.

"The human body can tolerate much more than we ever thought," Hubbard told me. "The most severe crash a driver has experienced in a HANS was when Richie Hearns experienced a front-to-rear chassis acceleration of 138 *g* with no head or neck injuries. He did break his foot, so we can't say he walked away from the crash."

3. First developed for automobiles in the 1960s, the air bag had considerable difficulty gaining public acceptance. General Motors introduced the air bag in its Oldsmobile Toronados in the midseventies, but the air bag failed to generate public interest. Front air bags remained optional until 1998 and even today remain controversial in their benefit.

   Each year, nearly two million air bag deployments occur among the more than eighty million air bags in use. In 2003, the NHTSA confirmed that there were 231 known deaths that resulted from air bag deployment in accidents that would have otherwise been nonfatal. A majority of the dead were women and children. While some estimates place the actual number of air bag–related fatalities much higher, the ethical argument against air bags remains unflinchingly clear, whether the number of deaths is higher or not.

Twenty percent of females are under sixty-two inches, the cutoff height for air bag safety effectiveness, making shorter females fifteen times more likely to die from air bag deployments than taller male counterparts. This and other gross discrepancies create a safety bias inherent in air bag technology—a bias that remains in general acceptance throughout the world. We wouldn't allow a pharmaceutical company to release a drug into the public with the same risks, yet air bag technology manages to avoid the same regulations. Recently, the problem of the vulnerable short-statured driver was allegedly resolved by Federal Motor Vehicle Safety Standard 208, but the solution isn't adequate, and it came at enormous expense. Put in terms of overall benefits, including costs, the simple seat belt arguably yields far more safety benefits than the air bag.

Sources

American Association of Railroads, Policy and Economics Department, *Class I Railroad Statistics* (Washington, D.C.: AAR, 2006).

Leonard Evans, *Airbag Benefits, Airbag Costs* (Leonard Evans, self-published, 2005).

Hubert Gramling and Robert Hubbard, *Development of an Airbag System for FIA Formula One and Comparison to the HANS Head and Neck Support* (Warrendale, Pa.: Society of Automotive Engineers, 2000).

AN INSULT TO THE BRAIN

Source

Floyd Skloot, "Gray Area: Thinking With a Damaged Brain," *LOST Magazine* no. 3 (Feb. 2006), www.lostmagazine.com.

ROB RABE CANNOT CRY

1. Tom Lutz, *Crying: The Natural and Cultural History of Tears* (New York: W. W. Norton, 1999).
2. In September of 2006, scientists at the Medical Research Council's Cognition and Brain Sciences Unit and the Division of Academic

Neurosurgery in Cambridge, England, and the University of Liège, Belgium, published their findings that an individual in a PVS demonstrated clear signs of willfull intention. Using functional magnetic resonance imaging (fMRI), the research team observed that the patient's brain indicated the ability to hear commands and respond to them in a cognitively appropriate manner. In a press release interview, team leader Dr. Adrian Owen commented:

> These are very exciting findings. This technique may allow us to identify which patients have some level of awareness. But it is important to emphasize that if we don't see responses in a patient it does not necessarily mean that they are not aware. Future work will investigate whether the technique can be used more widely in these patients and whether this discovery could lead to a way of communicating with some patients who may be aware, but unable to move or speak.

Source

A. M. Owen, H. R. Coleman, M. Boly, M. H. Davis, S. Laureys, and J. D. Pickard, "Detecting Awareness in the Vegetative State," *Science* 313 (2006): 1402.

PORTRAIT OF AN INJURY

1. Daniel F. Kelly and Donald P. Becker, "Advances in Management of Neurosurgical Trauma: USA and Canada," *World Journal of Surgery* 25 (2001): 1179–85.
2. Jam Ghajar, *Project Implements Severe Head Injury Guidelines in Eastern Europe: Study Will Help Measure Effectiveness of Guidelines*, American Association of Neurological Surgeons, www.neurosurgery.org.
3. Armando Basso, Ignacio Previgliano, J. M. Duarte, and N. Ferrari, "Advancement in Management of Neurosurgical Trauma in Different Continents," *World Journal of Surgery* 25 (2001): 1174–78.

THE RESURRECTION OF DOUG BEARDEN

1. The description of the HSV-1 virus pertains only to adults. The most common cause of herpes encephalitis in children is from the sexually transmitted HSV-2 virus, with infection usually occurring

at birth. Pediatric herpes encephalitis attacks the brain globally and does not show a predilection for any one area.

2. Voodoo and *voudoun* have the same meaning. Both operate as nouns and adjectives; I used the French spelling as the adjective.
3. Wade Davis, *The Serpent and the Rainbow: A Harvard Scientist's Astonishing Journey into the Secret Societies of Haitian Voodoo, Zombis, and Magic* (New York: Simon & Schuster, 1985).
4. Vail beds were recalled in 2005. While still in use in some facilities, Vail beds now require a strict protocol for safe use.
5. A savvy neurologist might have diagnosed Bearden with a case of Cotard's syndrome. Although Marcel Proust lampooned him as a gifted imbecile, the Parisian neurologist Jules Cotard made observations about the brain's plasticity that proved to be more than a hundred years ahead of his time. In the late nineteenth century, Cotard was the first to catalog the *délire des negations*, the delusion of negation. As a surgeon in one of France's most clinically rich psychiatric institutions, Cotard noted that an occasional psychotic patient would foster a delusion that their insides were missing, that their soul had been lost, or that they were dead. Cotard later theorized that the delusion was symptomatic of a severe depressive state. Later, psychology conferred the name Cotard's syndrome on patients who bore the delusion of death.

   Cotard's syndrome is categorized along with a host of other rare but disquieting reduplicative misidentification syndromes. In Capgras syndrome, an individual might believe that imposters have replaced all their loved ones. Fregoli's syndrome affects a person's ability to differentiate facial characteristics, so that every person appears to have the same face. The often-lauded neuroscientist V. S. Ramachandran theorized that Cotard's is actually an intensified version of Capgras, so that the person concludes they are dead because everything is unrecognizable or perceived as an imitation of a former life.

   While there is now little doubt that Cotard's is a brain-based disorder, some researchers suggest that biology does not fully explain Cotard's, and that there must necessarily be a peculiar mix of impaired perception and distorted reasoning that leads some people to conclude that they are dead. Whether the cause is neurological or neuropsychological, the fact remains that a cure for Cotard's continues to be elusive, and its resolution unpredictable.

ULTRAVIOLENT BRYAN

1. This time, Bryan entered the hospital with some resolve and some uneasiness. He asked his parents why he was the one who always had to have surgery, and he poked at his stuffed rat, TipToe. He drew his consolation from his mother and accepted his father's assurances. Under the fluorescent lights, the nurses shaved away Bryan's blond locks and then pushed an IV into his wrist. The surgery team propped Bryan's head right side up and began layering sterilization procedures that included covering nearby instruments. This would not be a delicate surgery.

   After the neurosurgeon peeled back Bryan's scalp and removed a sizable four-inch plate of skull, he discovered that the DNET had grown to the size of a golf ball, but he could not visually discern whether it had overtaken the entire lobe. Before the actual extraction of the tumor took place, the neurosurgeon needed a strong sense of what tissue was functioning and what tissue had been destroyed. Although it would add an hour to the surgery, the neurosurgeon conducted a brain mapping report that pinpointed the exact areas of damage.

   Intraoperative brain mapping involves the stimulation of specific portions of the outer brain using electrocorticography; deep brain tissue is too vulnerable to the same type of stimulation. At certain intervals, electrodes are placed directly onto brain tissue, and electrical activity is then measured and recorded. Motor and sensory mapping can occur while the patient is anesthetized, but a patient must be conscious for language mapping. The brain has no sensation, so the procedure is painless to the patient.

   "I remember waking up in surgery," Bryan told me. "It was strange because everything was blurry." They asked Bryan if he could feel his feet, and then if he could move them. He could.

   Through the display of voltage difference, the electrocorticography report revealed abnormalities between Bryan's frontal lobe and his temporal lobe. The results suggested that the tumor's damage was more widespread than anticipated. Based on the information, the neurosurgeon peered into the microscope hovering over Bryan's brain and announced he would proceed with a six-centimeter temporal lobectomy, including hippocampus. The measurement was precise and unforgiving: seven centimeters, and Bryan would no longer see anything in the upper field of his left eye.

2. Susan Curtiss, Stella de Bode, and Gary Mathern, "Spoken Language Outcomes After Hemispherectomy" *Brain and Language* 79 (2001): 379–96.
3. The Alex Center has since changed its name and location.
4. There are currently about forty-eight million children residing in the United States. The Centers for Disease Control reports that about 3 percent of children experience a brain injury each year, and that 75 percent of those injuries are mild. Crunch the numbers, and you can reckon that there are about three hundred and sixty thousand children in America with a moderate to severe brain injury. The number of students recognized as brain injured by the Department of Education is currently 14,844.

FUGUE OF THE PONY SOLDIER

Sources

Joseph Epes Brown, *The Sacred Pipe: Black Elk's Account of the Seven Rites of the Oglala Sioux* (New York: MJF Books, 1993).

Joseph Bruchac, *The Native American Sweat Lodge: History and Legends* (Freedom, Calif.: Crossing Press, 1993).

Anna Cantagallo, Luigi Grassi, and Sergio Della Sala, "Dissociative Disorder Following Traumatic Brain Injury," *Brain Injury* 13 (1999): 219–28.

Cherokee Nation Cultural Resource Center, *A Brief History of the Trail of Tears* (1998), www.cherokee.org.

John G. Neihardt, *Black Elk Speaks* (Lincoln: University of Nebraska Press, 2004).

IN ALL EARNESTNESS

1. There are numerous renderings of *The Gateless Gate*; for the purpose of this chapter, I have used examples from Koun Yamada's translation.

2. For an excellent starting point on the semantic complications regarding consciousness, take a look at "On the Neurophysiology of Consciousness" by the acclaimed researcher Joseph Bogen, in *Consciousness and Cognition* 4 (1995): 55–62 and 137–58.

THE HOSPITAL IN THE DESERT

1. As of early 2007, more than 23,000 U.S. military servicemembers have been wounded in Iraq and Afghanistan. An estimate from an echelon II medical unit places blast injury wounded at 88 percent, with 47 percent of those experiencing head injuries. While the Pentagon has yet to release hard numbers on brain-injured troops, citing security concerns, experts estimate the number of mild TBIs at over 7,600. About 2,500 brain-injured soldiers have already been treated through the Defense and Veterans Brain Injury Center, and the TBIs keep coming.
2. The Air Force Theater Hospital in Balad is staffed by 380 people. While an average Level I Trauma Center in the United States treats about 2,000 admissions a year, AFTH Balad admits 8,000. Eighty percent of patients at Balad require surgery, as opposed to 10 percent in the United States.
3. OR Procedures Chart from 333rd Expeditionary Medical Group Report, compiled by Air Force Theater Hospital in Balad.

WOOD OF THE SUICIDES

1. Among the brain-injured population, suicide is an issue that accompanies the injury. About 2 percent of brain injuries are caused by known suicide attempts, and more than 18 percent of all brain injury survivors attempt suicide in the five years following their injury.

Source

Grahame Simpson and Robyn Tate, "Suicidality After Traumatic Brain Injury: Demographic, Injury and Clinical Correlates," *Psychological Medicine* 32 (2002): 687–97.

ACKNOWLEDGMENTS

with the words going out like the cells of a brain,
with the cities growing over us like the earth
we are saying thank you faster and faster
with nobody listening we are saying thank you
we are saying thank you and waving
dark though it is.

—FROM "THANKS" BY W. S. MERWIN

After returning to work following an evaluation, I always stop by Dr. Rolf Gainer's office to discuss the case. A displaced New Yorker, Gainer has spent the last thirty years overseeing brain injury programs and studies in both the United States and Canada. Thousands of survivors live better lives because of his generosity and compassion, and he continues his work with quiet resolve. I spend hours listening to survivors and their families tell me about their lives, and Dr. Gainer in turn spends hours listening to me struggle with their cases. Dr. Gainer and I then bring the cases before a roundtable consisting of Ron Broughton, Juanita Edwards, Liz Lamers, Stephanie Maruska, Matt Maxey, Ken Pierce, and Pamela Washbourne. Behind closed doors, they are all heart. Together, the team at Brookhaven Hospital and its directors, Harvey Glasser and Steve Polkow, have managed to break open the doors of opportunity for a number of survivors, providing them with access to services they might never have found.

This book, along with my job, has taken me to nearly every state in America and into a number of homes. While traversing the country, I relied on the kindness of Gene Bolles, Milo Desmond, Jake and Natalie Dorn, Carol Elk, Marc and Allison Franklin, Don Greenfeather, Gary Graham, Brad Hendrix, David and Noël Kinsler, Dennis Leech, Bobby Mason, Todd and Sue Montgomery, Laura Napier, Alexis Sainz, and Brandon and Karen Scott. During my heartbreakingly brief stints in town, I regularly benefited from the benevolence of the Crouch families, Leslie Gainer, the Krabacher family, the Shadle family, and the Wenzels. Helen Cortes, Larry Epps, Ruben Habito, and David Williams provided me with their particular insights by phone and e-mail.

Little did I know that a brief conversation with VA psychologist Rose Collins would lead me to Iraq. My eventual trip to Balad Air Base was orchestrated through the help of airmen Lieutenant David Herndon, Captain John Upthegrove, and Dr. Eli Powell. Colonel Brian Masterson and Colonel Lorrie Cappellino, who facilitated my experience at Balad, are among the dedicated hundreds who have made the Air Force Theater Hospital at Balad a monument to military medical history. Their work has served to advance the acute treatment for all future brain injury survivors. In Balad, and throughout Germany, the generosities and collective levity of Brent Henderson, Kenny Liston, Sergeant Scott Reed, and Joel Schwartzberg proved to be soul-nourishing company.

A book, I have learned, is no small miracle of efforts by a motley band of individuals. Over cold drinks on a hot Tulsa day, Anne Reid Garrett and Jim Fitzgerald asked me about my job, and then suggested that I write about it. Anne, now my literary agent, did everything right. She found Paul Elie at Farrar, Straus and Giroux, who saw in the book a story that demanded telling. I am grateful for Paul's sincere commitment to the book, as well as his intuitive guidance throughout the writing process.

My start as a writer began the day I received a term paper dripping with red ink from Ruth Weston; she, along with Bill Epperson, Linda Gray, and Grady Walker, taught me the weight of words. Through his careful and attentive reading, Darren Ingram perpetuated their work. I owe special thanks to Bob Guccione, Jr., Patti Adroft, Corey Powell, and the rest of the crew at *Discover* magazine for their support and insight. Suzanne Wallis, patron saint of Tulsa writers, was instrumental in this book's genesis. The very core and being of this book, however, is borne upon a sacrifice of the dearest kind.

At fifteen, I told my mother I wanted to marry a girl named Christy. My mother, Caridad Palmón, rolled her eyes and later bought me a laptop as consolation for the lonely years ahead. My first short story was about Christy, of course; her presence can be felt in everything I've written since. At twenty-seven, I married her, and we now have two daughters together, Cherish and Amaya. My mother and Christy both endured countless unwashed dishes and dirty diapers, and offered generous helpings, while my girls offered impromptu performances and kisses on the way out the door. You each sustained me through thousands of lone miles, and now, look, the book is finished! We made it.

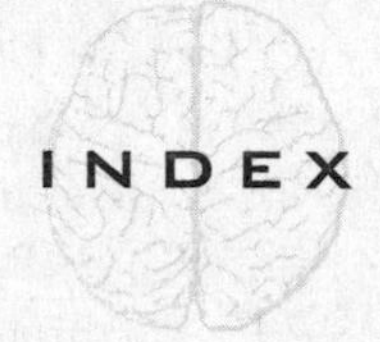

INDEX